Iterative Product Development:

How to Turn Your Ideas into Successful Products

Alan Lee Starner

About the author

Alan Lee Starner has been developing products since the early 1990's. He currently has 7 patents under his name and has brought several of these products to the marketplace. He earned his Bachelor of Science in Physics with a minor in Computer Science in 1981. In his early career he worked in the defense industry later, transitioning to being the director of software for a semi-conductor equipment maker.

His greatest education in product development happened in the University of hard knocks. He wrote this book to help other product developers avoid the painful mistakes he has made along the way. You can reach him at alan@eafproducts.com, or go to

www.nalalab.com

Acknowledgements

Special thanks to Kai Madrone for her support, encouragement and editing assistance. This book would not be as readable without her help.

CONTENTS

Introduction

Turning your idea into a successful product is exciting and, if done well, can give you the kind of financial success you've only dreamed of. However, it is also high risk. Estimates are that 70% to 95% of new products fail soon after being introduced and this doesn't count products that never even made it on the market. You obviously want to be among the 10% or so of products that do make it, but how do you increase the odds in your favor?

The short answer is to have a disciplined method that will keep you focused on the most important tasks, prevent you from skipping steps, and can help you know when to hire someone for a given task and when to do it yourself. In bringing a product to market, there are so many interrelated fields of expertise that it is really easy to get off track, so unless you have a cross-functional team already in place with a wide range of expertise, you will likely need some guidance along the way.

I started doing product development in 1987 and have had both failures and successes. Though success feels great, I've learned far more from my failures. It's hard to forget the sting of losing the time and money invested into a product idea that doesn't work out, especially when 20/20 hindsight makes the mistakes obvious. You can study design, you can study marketing, and you can study manufacturing, but to create a successful product you need to incorporate all of these and more. Most of the engineers, designers and marketers you meet will have only done their slice of the pie: either as an employee or a consultant with contracts in place. With no skin in the game, their risk is limited. It feels completely different when you are the entrepreneur taking the risks and spending the money – you have *nothing but* skin in the game. There aren't many people who have gone through the entire product process from conception to sales, and most of those who have are not in the business of helping others get their product to market. That's where this guide comes in.

I would like to share an approach to product development which is both disciplined and holistic - it integrates all the slices of the pie into a unified approach. I would like to help you avoid the many mistakes that I've made and offer you a focused step-by-step way to develop products that hit the target for your customers. I recommend reading through the entire guide before taking action because many of the elements are interrelated and understanding the entire journey will help you with what you are doing each step along the way.

Although I developed this approach independently, it has much in common with lean product development and agile software development. It minimizes costs and risk by doing as much work as possible upfront, emphasizes a fast iteration cycle combined with customer feedback, and has a relentless focus on the customer's needs. However, for physical products that involve expensive tooling before mass production, the agile principles apply mainly to the early phases where 3D printing prototypes substitute for the release of new software versions. With physical product manufacturing, this quick iteration cycle should happen before expensive tooling is ordered since changes from that point forward are expensive and create huge delays.

There are multiple roles required to bring a product to market, and that means either having someone on your team who specializes in each role or wearing a lot of hats. You can wear as many hats as you like, but since success depends on each role being well fulfilled, you should only play a role in which you have competence.

Though there are many different roles to consider, I consider the following six to be the most essential roles for the product development process. Once the product is on the market, additional roles come into play.

The six most essential roles in the development process are:

Product Champion: This is usually the inventor or entrepreneur who came up with the idea. The product champion is responsible for holding the vision of the product, making sure the whole process stays on track, and that the final product fulfills their vision of customer satisfaction. The vision they hold should be of the customer being satisfied with the product, not a vision for the product itself. This allows our ideas about the product to evolve in ways beyond our original conception and not get stuck in our first set of ideas. Many new products fail because the focus was on bringing the original vision to life, rather than having a relentless focus on the customer's needs and desires. The Segway is a well-known example.

Product Designer: This is usually someone with design and manufacturing skills. The Product Designer will work with the Product Champion as an ally to move the product forward. These days most manufacturers will want a 3D model, so if your product requires that, look for someone using a modern 3D Computer Aided Design (CAD) package. A good designer will also help uphold the vision of the product to make sure it satisfies the customer's needs. If the product champion also has product design skills, then these two roles can be combined together, though it often helps to get additional viewpoints with design decisions. It can also work well if there is a team of people who play both roles.

Customers: These are the people who will eventually want to buy your product and, in many ways, play the most important role. You can't have success without pleasing your customers. You can and should use representatives in this process in the beginning, but it is extremely helpful to have actual potential customers involved in the process as well.

Suppliers: Suppliers are an important source of feedback on manufacturability and costs. They become more important as the process gets to the later stages, and we may end up being used during the manufacturing phase.

Sales and marketing: If you have sales and marketing people in your company, they can be good sources of feedback. They are the ones who will be explaining the product and its benefits to customers and can have valuable insight into what will help things sell well. They may be the best people within your company to represent the customer's viewpoint.

Chief Executive Officer (CEO): No, you don't need someone with the title of CEO, but your team will need someone to oversee the product development process, keep things on time and on budget, and make sure nothing falls through the cracks. You will need someone to make sure the business plan gets executed and executed well.

There may be others playing roles as well. If you were going to climb Mount Everest, you would probably hire a guide – someone who has been there before and lived to tell the tale; someone who knows what it's like to pour your heart, soul, time and money into a project. If you are developing a product, you have skin in the game; if you have a mentor who has been through the product development process, they can help you avoid the pitfalls and work more effectively toward your goals. If at all possible, bring someone onto your team that has experience with developing a product and bringing it to the market.

Managing risk is critical if you want to be successful. The risks in the beginning while you are sketching your idea on the back of a napkin are small, and these risks grow each step along the way. In the next section we break down the different phases of product development and how to know when you are done with one phase and ready to go on to the next.

Product Development Phases Overview

I divide product development into four sequential phases of activity, with each phase being more expensive (thus higher risk) than the phase that preceded it. Here is a quick summary of the four phases - we'll dive into each phase in detail later:

Phase 1 - Ideas and High-Level Estimates (Almost Free):

This phase is a quick check on the idea and whether it is viable. This phase has very few costs other than your own labor. If you're like me, the thought of creating your own product will fuel you through this process. Front load as much work as possible into this phase to reduce your risk. There is little financial risk in this phase. What you are risking is your time. Expect to spend 40 to 60 hours or more in this phase, even for a simple product. If you decide not to go forward after your evaluation, you will know that at least you investigated the possibilities. Rejecting product ideas that don't meet your risk/benefit evaluation is part of a good product development practice and develops evaluation skills for the next idea that comes along.

Phase 2 - Iterative Product Development (Moderate Cost):

The task of this phase is to design the product, make prototypes and then test those prototypes with potential customers. Based on feedback from the testing, the design is updated, more prototypes are made, and we test again. This design-make-test cycle is iterated until we have the design nailed – not just in our minds but in the experience of the potential customers.

Costs are a bit higher, but modern technology like 3D printing and home Computer Numerical Control (CNC) machines will often allow the creation of functional prototypes from your designs for minimal costs. If you are handy, or have talented friends, you may be able to put together prototypes yourself. That allows you to put your product in front of customers during product development to get their feedback. This is super valuable and should be a part of product development whenever possible. Changes to the design are fast and relatively easy. Getting the product wrong here is no problem because you can just iterate the design when problems are found. The money here is primarily at risk if we get our High-Level Estimates wrong since we can easily pivot the design if we run into problems. Each design iteration has some risk, but the cost/risk of each iteration is low.

The biggest risk in the Iterative Product Development phase is stopping before you have the design nailed, or not getting objective independent feedback (friends and family tend to give glowingly positive or in some cases overly negative feedback). The purpose here is to have a design that has been verified by customers to meet the market need. For a simple product, this phase can cost several thousand dollars for paying product designers and creating prototypes, but this cost is spread out over multiple iterations of the design. More complex products will take more time and money to develop, as will starting with only vague ideas about what your product will look like. Fail fast and fail often until you have your product design nailed.

Once we have our design nailed, we go into the pre-production phase where we choose our suppliers and update all of our estimates.

Phase 3 - Mass Production (High Cost):

Once a manufacturer starts developing tooling for mass production, the costs go up significantly. This makes design changes difficult or impossible, and when possible, it's always expensive. If you have done an excellent job in the previous phase and have good prototypes and feedback that you have nailed the market need, then the main risks are supplier problems. Make your mistakes in an earlier phase if at all possible. You don't want to do your initial functional testing with mass production parts.

Phase 4 - Selling and Support

(Very High Cost):

This might not be considered product development by some but should be included in the development process since this is where we are headed and the place where the value of the previous phases is realized. Unless you want to develop a product as a hobby and don't plan to sell it, considerations about this phase should be part of the context of all other phases.

Your idea can die here as easily as anywhere else, so your plan needs to bring you to the point where you are selling and supporting your product. Product mistakes that make it into customer hands are the most expensive of all. Not only have you already paid for the tooling and initial production run, but there is also a high cost to your reputation with customers and you will spend a lot of time and money on customer support or just end up going out of business. Once customer trust is lost, it is very hard to regain it – just like in any other relationship.

The phases of product development are shown graphically below:

Product Development Process

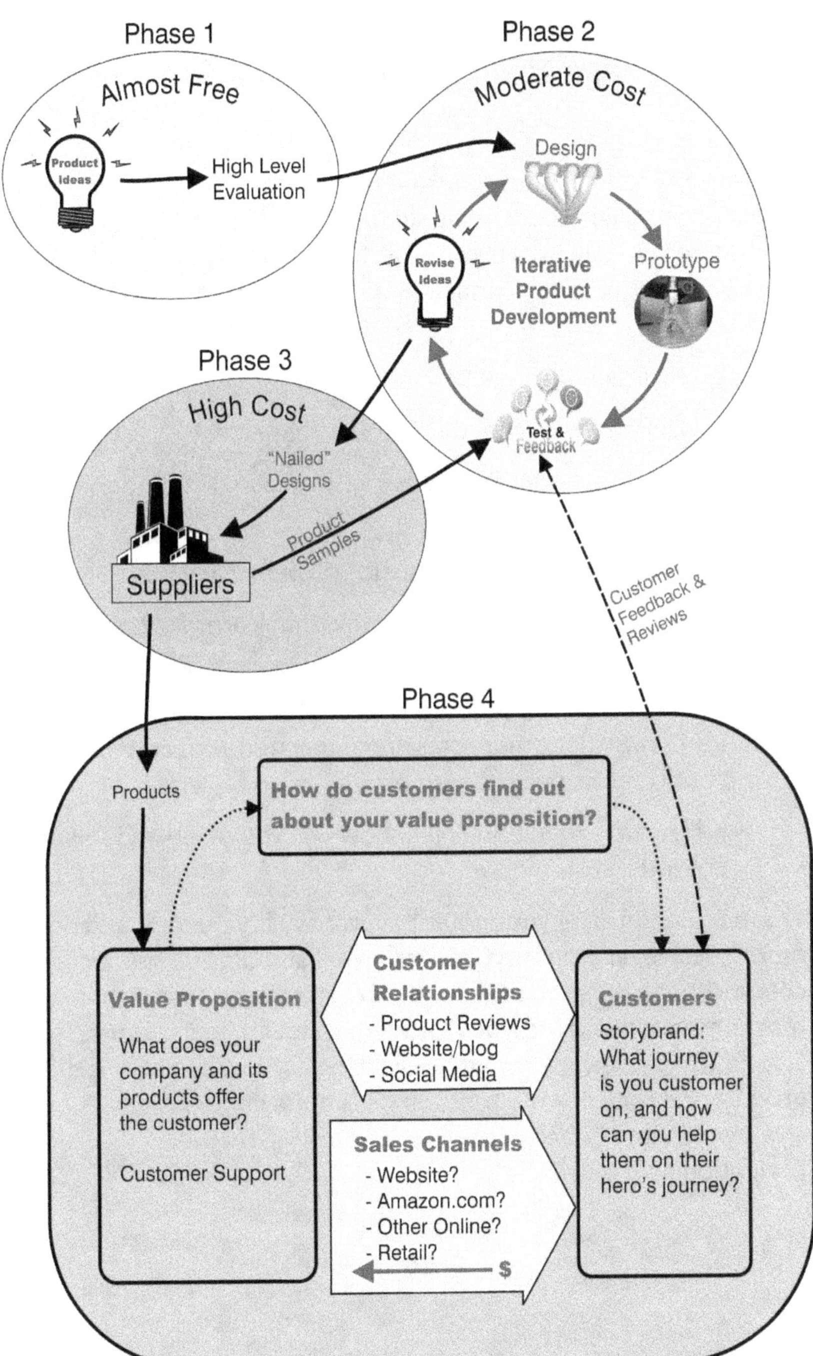

Phase 1: High-Level Evaluation

Before you spend a lot of time and money designing and manufacturing a product, it would be good to know how many you can reasonably expect to sell at what profit margin, and what your distribution channels will be. To calculate this, you need to:

[_] Estimate the production cost

[_] Define your initial distribution channels

[_] Define a target sales price within each distribution channel

[_] Understand how customers will perceive your product

[_] Understand market size – how many are likely to be purchased each month? Is it seasonal? Through what sales channels?

[_] Know what other costs are associated with getting your product out and supported

[_] Define a marketing plan: how will you reach and sell to potential customers?

These will determine your potential profitability per unit and per month. However, you can't get accurate estimates on production cost without having a design, and you can't gauge customer reactions without prototypes for them to touch, feel and use. Should we just rely on our gut instincts, plow forward with the product and hope for the best? Hope is not a good business strategy. There is a better way, which I call High-Level Evaluation.

The High-Level Evaluation is the first phase of product development. The costs can be extremely low if the work is done yourself. You can hire someone to help you, but I recommend you do it yourself or at least stay really close to the research so that you have first-hand knowledge of the information provided. The evaluation does not have to be perfect; a rough estimate will tell you if it's worth spending additional time and money developing the product; many of my own product idea have died right here – and that's a good thing. Later, a more detailed analysis is appropriate before you pull the trigger by spending the big bucks on manufacturing. By the time you get there, you will have collected a treasure trove of information about your product and how it fits into the marketplace. Right now, we just want to do our estimates without spending any (or very little) money; a sort of pre-check to see if further effort is warranted.

Here are the items to estimate or develop:

> **Production Cost:** One way to do a quick product cost estimate is to find similar items for sale on Alibaba or another sourcing web site. You'll have to do your best guess on how your design will differ in cost from what you find there. Is your design more complicated? Does it use similar materials? If your product is completely new and there is nothing like it anywhere, the problem is more challenging, but looking at products in the same industry made of the same materials and of similar size and complexity can give you a rough idea.

> Many products require tooling or injection molds. These are one-time non-recurring costs but should be taken into account to determine profitability. A simple plastic

injection mold can be $500, while a complex mold can be tens of thousands of dollars, or even more. A water bottle lid I recently created required 5 molds plus an overmold, and the total cost for these injection molds was about $10,000 in China. Also consider what quantity you would initially order. Many companies have a minimum order quantity, and this can be in the thousands. The higher your startup costs, the higher the risk.

Where will your product be manufactured? It is important to include shipping costs as part of your product cost estimates. A 20-foot shipping container from China to the US is around $3000 (in 2019), and that is just from port to port; there may be additional charges for train/truck to get it to your location. It also doesn't include import duties, and these can be significant. At the time of this writing, there is an additional import duty of 25% on most goods manufactured in China. Shipping costs go up and down all the time, so the above is only a very rough estimate and costs can vary widely with market conditions.

You may also have different costs with different sales channels. For example, if you sell on eBay or your own website, you may have the cost of a box, labels, and packing materials in addition to postage/shipping to your customer, plus any selling fees or credit card processing fees.

The production costs determined here are just our first best guess. We will update these with more accurate estimates once we have the design done and can get quotes from suppliers. Still, we want to be as accurate as possible.

Initial Distribution Channels: How will your customers get the product? There are online options as well as brick and mortar options. You don't have to have all possible channels defined, just the ones you will start with. Other distribution channels can be added later.

Half of all online retail in the US is through Amazon, so if you plan on using Amazon check your costs there. I can tell you that Amazon charges a 15% commission for most products, and additional fees if you send to their warehouse for Amazon to fulfill (Prime), and the additional fees are based on the size and weight of your product. Some other online options are E-bay and your own website. Walmart is also allowing select vendors to sell through their platform.

For offline sales, determine what types of stores your product would best fit and determine the costs/markups that are typical for your product. Cheaper products usually have a 100% markup: if you sell it wholesale for $5, the store will probably mark it up to $10 for their customers.

Price: This is probably the easiest to estimate. Find your competitors on Amazon or other websites and note the price they are selling for. I recommend creating a detailed spreadsheet with each competitor and the price. Don't forget to factor in shipping, especially if your product is a heavy one. For some products, customers expect free shipping.

If you plan on selling in retail stores, visit a few likely

candidate stores and notice the pricing of similar products as well as product placement. You might ask a salesperson about your competitors, how they sell and what people like and don't like about them. Online price evaluation is best done while doing the Market Perception Evaluation as noted below. Based on the features and benefits of your product versus your competitors, determine a target retail price for your product. You can then estimate wholesale prices for your product category based on the retail price target.

For example, suppose that other similar products retail for $14.95. If your manufacturing cost is around $2.00 and the retail stores double that price to get to their retail price, that would support a wholesale price around $7.50. If your manufacturing cost is higher, say $4.50, you might need to price it a little higher to make a profit.

Market Perception: This is the hardest to estimate. Presumably your potential product has some clear benefit(s) over what is currently available. I suggest you try to gauge the pain points of your competitors, and how strongly your customers feel about these. Go to Amazon and other websites with product reviews and spend a lot of time reading your competitors' reviews. I suggest keeping detailed notes about what features customers like, as well as what they don't like. This is best done in the same online sweep where you find the prices. There may be a tiered price structure with upgraded feature sets being at a higher price tier, so note how your product would fit into any tiers. Cars with V8 engines generally cost more than cars with 4-cylinder engines.

This information will also feed into the product design: you want your product to have all the things customers like, and avoid all the pain points of your competitors, without introducing new pain points of course, and without being too expensive. Don't skimp on this step. You should be prepared to spend 20 – 30 hours or more doing this, even for a simple product. Each time I have done this I found important features that affected my product offering. For example, I sell a water bottle straw lid and in the reviews of competitors that offered a straw cleaning brush with their straw lid, customers raved about the included brush, often barely mentioning the lid. It's such a simple low-cost addition that I immediately added it to my product. If you have access to sales and marketing people, I suggest you run your ideas past them for feedback as well.

Market Size: This is another evaluation that can be done in the same online sweep where you research Price and Market Acceptance. On Amazon, they rate each product with a Best Seller Rank (BSR), which you can see in the screen shot below:

The Amazon Best Sellers Rank, or BSR, is a ranking of each product within their major categories. The above item is the 6th bestselling item in Amazon's Sports and Outdoors category, and it's the second bestselling item in the water bottle sub-category. The lower the BSR, the higher the sales. Note that Amazon updates these rankings every 2 hours, so what you are getting is just a snapshot in time.

Additional Information

ASIN	B09BZ4DXC3
Customer Reviews	4.8 ★★★★★ ˅ 45,923 ratings 4.8 out of 5 stars
Best Sellers Rank	#6 in Sports & Outdoors (See Top 100 in Sports & Outdoors) #2 in Water Bottles
Date First Available	August 5, 2021

You can use the rank within the sub-category to help you find other competitors' products, and you can use the rank within the major category to get an estimate of that product's unit sales per month.

The rankings don't correspond linearly with sales; a product with a BSR of 50 might sell 4 times the units of a product with a BSR of 100. The tool I use to convert BSR to monthly sales is Jungle Scout's sales estimator, which is currently available for free. Go to https://www.junglescout.com/estimator/ and enter the BSR (6 above), product category (Sports and Outdoors above), and the marketplace. Since I sell in the US, I just use the US marketplace, but if you have plans to sell elsewhere, you can get estimates there as well. As I write this, the estimated sales per month for the above product in Sports and Outdoors is 22,740 units per month.

Get the estimated sales for each of your competitors and total them up. This is an estimate of the maximum possible sales on Amazon were you able to capture every single customer for that product. Amazon is about 50% of online retail, and online retail is estimated to be about 10% of total retail and trending upward (2018 data). This percentage online likely differs based on the type of product you are evaluating, as some products are mostly sold retail, and others online, so use this information as a general guideline with considerable uncertainty.

There are industry groups that also track sales within their particular industry, and I suggest you do some web searches to see if you can find any interesting information that can help you gauge the potential market for your product.

Do your best to estimate the potential sales per month and if it's a seasonal product, like camping gear, then you will want to estimate at the market max and minimum, as well as the 4th quarter sales spurt. Almost all products sell more in the 4th quarter with holiday buying and it's common for half of annual sales to happen in the 4th quarter.

Percent of Market: What percent of the market can you reasonably expect? That depends on many factors, but I like to see it being worth my time with minimal market penetration. If I can make it work at 1% market share, then I will be really happy at 10% or more. I suggest doing an optimistic estimate as well as a 'low ball' estimate (the minimum market penetration you think you can achieve).

Admittedly, this is highly subjective and generating an accurate estimate will be error prone. If your product provides no new benefits to the customer, then the percentage is likely to be quite low. Without too much wishful thinking, do your best to generate a realistic estimate and it can be refined later when we start getting feedback from customers.

Other Costs: Be sure to include other costs that you will have in selling your product. The following is a partial list:

[_] Fulfillment costs: postage, shipping supplies

[_] Amazon fees (if selling on their platform)

[_] Warehouse/storage costs

[_] Packaging and/or labor if applicable

[_] Employees, Insurance and other overhead costs

[_] Taxes

[_] Educational costs – if your product needs to be explained, factor this in

Marketing Plan: It's not enough to have a great product for a great price. If nobody is aware of your great product, you won't sell any. Part of your initial evaluation should include a strategy for how you will get your product in front of customers and give them a chance to buy it. There are a lot of marketing strategies around and I recommend you figure out who your customers are and how they spend their time in order to get some insight into how to connect with them. Here are some possibilities for getting the word out:

Influencers: Find some review websites/bloggers that target your audience. Research ways that you might reach out and offer free products to influencers. For example, many are Amazon Affiliates, so they make money by linking from products on their websites to Amazon. If a buyer gets to Amazon that way, the influencer will get a small commission from anything they buy while on Amazon, even if it's not the product linked to. Some influencers are "pay to play," and will review your product for a fee.

Social groups: Look for user forums, Facebook groups, magazines, and other content that target your audience. These can be a good place to advertise and/or a good place to offer free products (prototypes) in exchange for their opinion. Getting user feedback on your designs is a critical component of product development (more on this later).

Advertising: Advertising is the brute force but expensive method to reach customers. It can be done effectively, and it can also be done poorly where your sales increase doesn't even pay your advertising costs. Advertising costs are not generally included when determining gross profit margin but will impact on your net profits. Some people budget a fixed percentage of sales for advertising. If you are not sure, I suggest allocating 10% of your sales to advertising and marketing. There are a lot of products out there and most people have a high barrier to looking at a new product. Getting your buyers' attention can cost money. If you plan on getting traffic to your website using Google Ads, investigate what keywords would be effective and the cost per click for those keywords. Amazon has pay-per-click ads that function similarly. Advertising on Facebook or YouTube is another possibility.

Content Marketing: What articles can be written to underscore the importance of the design features of your product and how it meets important needs? How can that information be shared so that potential customers see it? For example, if you sell a selfie stick, you could write an article on how to get the best shot and what features (that your product happens to have) are essential for those great shots.

Putting it all together

Once you have all of your estimates done and your high-level plan put together it's time to decide if it's worth going into the design phase. Let's say we have completed our research on Widget X, and competitors are selling in the range $13 - $18. We know our product is better because of the great new features we have planned, but to be on the safer side, we have decided on a target retail price of just below the middle of that range: $15. Let's say our estimated production costs are as follows:

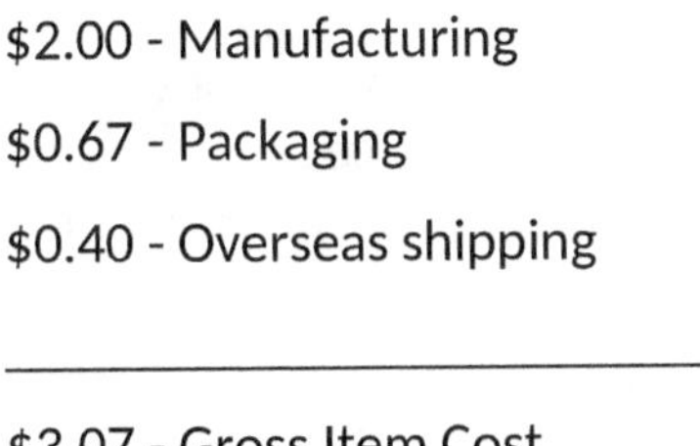

$2.00 - Manufacturing

$0.67 - Packaging

$0.40 - Overseas shipping

____

$3.07 - Gross Item Cost

The item has non-recurring costs for a plastic injection mold and other tooling totaling $10,500, and we estimate the design and prototyping to be $4000, for a total non-recurring cost of $14,500. Our overhead costs are low at only $700/month since we will start this "in our garage." A simple summary of profitability might look like this:

Estimates by Sales Channel	Market Size Units/mo	Sales Price	Channel selling cost - Percent	Fulfillment	Net item cost	Net Profit per item	Pessimistic			Optimistic		
							Market share	Units	Net Profit	Market share	Units	Net Profit
Amazon	15000	$15.00	15%	$3.75	$3.07	$5.93	1%	150	$190	15%	2250	$12,643
Wholesale	100000	$8.00	5%	$0.00	$3.47	$4.53	1%	1000	$3,830	5%	5000	$21,950
TOTAL								1150	$4,020		7250	$34,593
Months to payback non-recurring									3.61			0.42

Notes on the spreadsheet:

The sales price can be different for every sales channel. When selling lower priced items wholesale, the retailer will often come close to doubling the wholesale price to get its selling price.

Each sales channel may have different costs. In the above example, we have a sales commission of 15% of the sales price for selling on Amazon. In addition, there is the fulfillment fees of $3.75, plus the manufacturing cost of the items itself.

Net Profit is sales price, minus our channel sales cost, minus the cost of manufacturing the product.

Please note the final row in the above chart that indicates the number of months at that profit required to pay back the non-recurring start-up costs.

No one can tell you what level of profitability is required for a given product. For me, a lot depends on how confident I am in my estimates and the level of risk. Remember at this point that we are just doing the High-Level Evaluation and we will have an opportunity to refine our estimates later before a final decision to manufacture.

Let's take a brief pause to look at how the risk level for the above project looks in the product development, manufacturing and sales & support phases that follow. The analysis below does not include the hours spent by the Product Champion, so if you include that, the results would be even more dramatic.

Iterative Product Development: Financial risk level from the example: Up to $4000 for the phase, but much less for each product iteration, let's say 10 iterations at $400 average cost per iteration – that's a lot of learning. In practice, the first iteration may be half of the total cost, with subsequent iterations being 5% - 10% of the total cost.

Mass Production: All the non-recurring costs are at risk in this phase. This includes the $4000 from the Iterative Product Development phase, plus the $10,500 tooling costs and production costs. In this example, assuming an initial production run of 10,000 units, the production costs are $30,700. The total amount at risk is $45,200.

Selling and Support: Financial risks are harder to quantify at this stage as it includes intangibles like the risk to your reputation with customers. Tangible costs include all the costs of the previous phases, as well as all selling costs and overhead costs.

The bar chart below graphically illustrates why we want to do as much learning in the earlier phases in order to reduce financial risk. It shows how much it costs to fix a mistake in each product development phase.

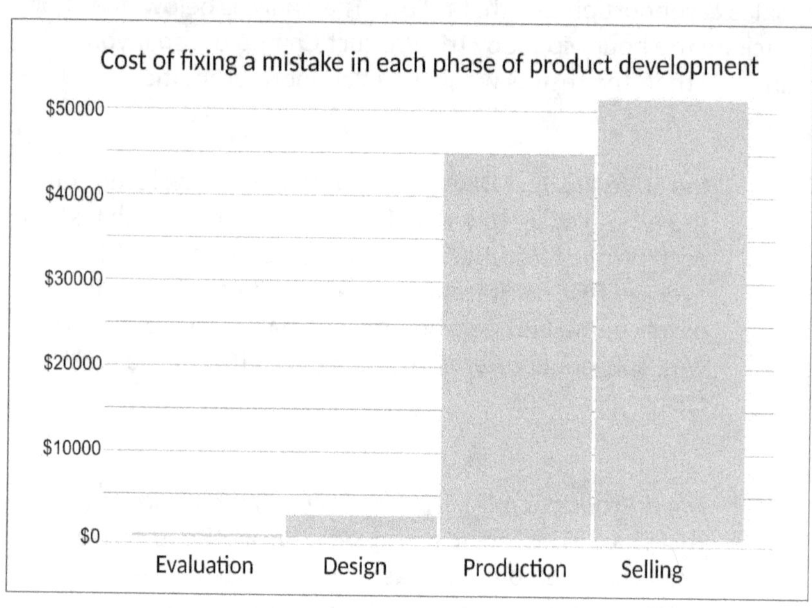

Besides getting our cost estimates, upon which we can base a go/no go decision if we've done our job well, we have also collected a lot of data on what customers in our market niche want and what their pain points are with current product offerings. This information will be extremely valuable in the next phase where we design, produce, and test prototypes.

Buy your competitors products

But before we start working on design, there is another thing I recommend: buy a sample of each significant competitor's products. Reading reviews and looking at feature/benefit lists is good, but there is no substitute for being able to touch, feel, and use the actual products. Hands-on contact with your competitor's products will put all that good information you found into context and provide grounding for your ideas. The buying can be done integral to your investigations (i.e., before you read the reviews for that product), or you can do it toward the end of the phase. If you do it at the end, be sure to update your analysis with anything you learn before making a decision to move forward.

Hold 'em or Fold 'em?

With the High-Level Evaluation almost complete, it's time to ask the 1-million-dollar question: Should you move forward into the Iterative Product Development phase? You know what features you want the product to have to be competitive. You've identified your distribution channels, price points, and a plan for how to reach customers. You have a market analysis and cost analysis, and a grounded idea about your profit potential per month and an estimate of the costs to get it into production and support sales.

The High-Level Evaluation ends with a summary of all the hard work you have been putting into it and doing the risk/benefit analysis. Is the potential benefit worth the risk? You will be risking your time as well as money, and the profit needs to be worth the risk. This is the first GO/NO GO determination we will make. Once we enter the following phase, the Iterative Product Development phase, we refine all these estimates and at the end of that phase, we will decide once again whether to move forward.

Our decision now is whether things look promising enough to invest time and money into the Iterative Product Development phase, where the costs go up. If it does not look promising now, it is unlikely to improve in the next phase, and it might be wise to look for a better business opportunity.

High-Level Evaluation Checklist

[_] Gain a good understanding of how customers will likely perceive your product

[_] Create a detailed estimate the cost to manufacture, assemble, package and ship your product to your distribution channels

[_] Define initial distribution channels and any associated costs

[_] Identify a target sales price within each distribution channel

[_] Create sales projections for each distribution channel including costs and profits with at least a pessimistic and an optimistic sales assumption

[_] Develop a good understanding of your products' market size and how customers are buying similar products

[_] Estimate other costs associated with getting your product out and supported

[_] Create a marketing plan for reaching potential customers

Phase 2: Iterative Product Development

In this phase we will develop, refine and test prototypes that satisfy our customers. For most people, they will need to hire someone to create the 3D CAD files or other inputs that will be required to mass produce your product. All the research we did on our competitors in the previous phase will be extremely valuable here as we design a product that best meets the needs of the market.

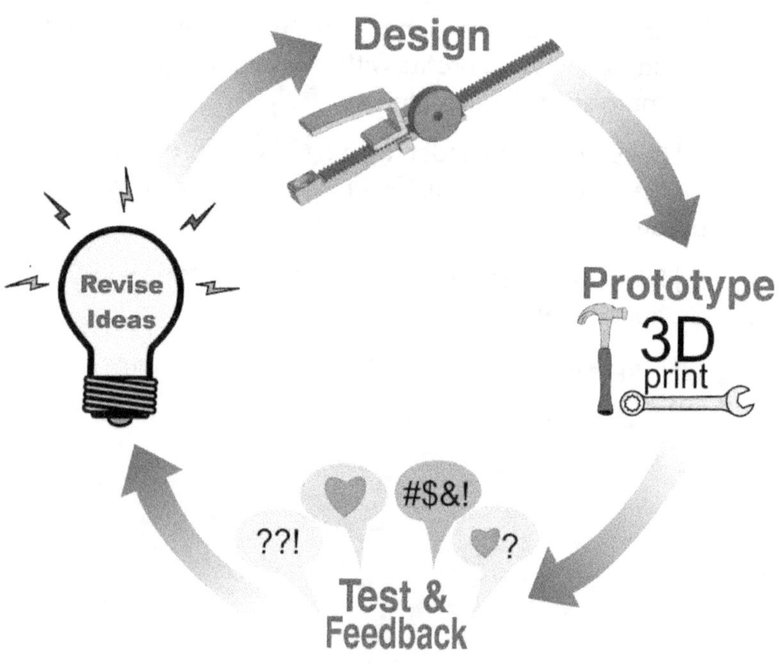

Overview

There are four sub-phases that are repeated iteratively until the product design fully meets the customer's needs:

Design: Create designs for the part(s) of a product that will meet the customer's needs as currently understood. The first time through the cycle these needs come from the original idea plus the results of our research from the High-Level Evaluation phase.

Prototype: Functional prototypes are created based on the current design.

Test and Feedback: The prototypes are given to customers or the best representation of the customer that we can find. They will test the prototypes and give feedback on what they like and don't like.

Revise ideas: The ideas that the product is based on will be revised to incorporate the customer feedback. These ideas will be used to modify the design for the next iteration of the cycle. When there are no more improvements to be made, or if those improvements fail a cost benefit analysis, then we might be ready to conclude the Iterative Product Development phase.

The Iterative Product Development cycle will continue until the design is mature enough to meet the market's need. It can take many cycles for this to happen. The first few iterations may involve testing and feedback with a small number of people – perhaps as few as the Product Champion (the one who created or is championing the product idea) and the Designer(s). Once they are happy with the prototypes, it is time to get more objective feedback from customers and/or their representatives and continue the cycle a few more times.

Let's take a deeper dive into each of the sub-phases:

Design

In the design sub-phase, we will take all of the information we have so far and create a design that meets the customer's needs. This is primarily the job of the Product Champion and the Product Designer. It's the job of the Product Champion to hold the vision for the product and make sure the design will meet the customer needs. It's the job of the Product Designer to help that process along, while also keeping an eye on manufacturability and costs.

There's a lot that goes into good design. The technical aspects of how to create parts for injection molding and other manufacturing technology are beyond the scope of this book. Instead, I want to focus on what makes for good design from the non-technical perspective of the average Product Champion. We all use and buy products and it's at this user level that the design must satisfy the customer. You probably recognize good design when you see it.

The iconic product designer for Braun, Dieter Rams, introduced the idea of sustainable development and of obsolescence being a crime in design in the 1970s. Accordingly, he asked himself the question: is my design good design? The answer he formed is now the celebrated ten principles of good product design, shown below.

Dieter Ram's Ten Principles of Good Design:

Good Design:

1) Is innovative - The possibilities for progression are not, by any means, exhausted. Technological development is always offering new opportunities for original designs. But imaginative design always develops in tandem with improving technology, and can never be an end in itself.

2) Makes a product useful - A product is bought to be used. It has to satisfy not only functional, but also psychological and aesthetic criteria. Good design emphasizes the usefulness of a product whilst disregarding anything that could detract from it.

3) Is aesthetic - The aesthetic quality of a product is integral to its usefulness because products are used every day and have an effect on people and their well-being. Only well-executed objects can be beautiful.

4) Makes a product understandable - It clarifies the product's structure. Better still, it can make the product clearly express its function by making use of the user's intuition. At best, it is self-explanatory.

5) Is unobtrusive - Products fulfilling a purpose are like tools. They are neither decorative objects nor works of art. Their design should therefore be both neutral and restrained, to leave room for the user's self-expression.

6) Is honest - It does not make a product appear more innovative, powerful or valuable than it really is. It does not attempt to manipulate the consumer with promises that cannot be kept.

7) Is long-lasting - It avoids being fashionable and therefore never appears antiquated. Unlike fashionable design, it lasts many years – even in today's throwaway society.

8) Is thorough down to the last detail - Nothing must be arbitrary or left to chance. Care and accuracy in the design process show respect towards the consumer.

9) Is environmentally friendly - *Design makes an important contribution to the preservation of the environment. It conserves resources and minimizes physical and visual pollution throughout the life cycle of the product.*

10) Is as little design as possible - Less, but better – because it concentrates on the essential aspects, and the products are not burdened with non-essentials. Back to purity, back to simplicity.

I mostly agree with his principles. The one area I'm not in complete alignment with is principle 5. Sometimes a product is a work of art, and sometimes it should be decorative. He also says in principle 3 that products should be aesthetic, so perhaps I am splitting hairs here. He has one overriding principle which is even more important than these ten:\

> *"Indifference towards people and the reality in which they live is actually the one and only cardinal sin in design."*

I could not agree more.

Customer Delight Questions

Expanding on Dieter Ram's final quote, I have developed some questions I suggest you ask yourself and your team about your product and how it fits into people's lives:

[_] How and when will they use your product?

[_] Where will your product be stored when not in use, and how much space will it occupy? Does it want to be in a case or storage box?

[_] Does your product need to be cleaned, and how is that done? Is there anything that can be done to make this easier or give them a reminder? Is there anything you should include in your product to facilitate cleaning? For example, most new prescription glasses come with a special cloth to clean the lenses.

[_] How does the size and portability of your product compare to your competition?

[_] Are there any hassles involved with using your product, and is there anything you can do to eliminate or reduce them?

[_] Is there any setup or calibration required for your product? If so, how can you make this as easy and fast as possible?

[_] Does your product need to come with instructions? If so, consider the instructions as part of your product. Make them as simple and easy to understand as possible. Also make them graphical as much as possible to avoid any problems for those who speak a different language. The instructions should be given to your customer representatives in the test phase.

[_] Is there periodic maintenance required, and is there any way to reduce, eliminate, or simplify that maintenance?

[_] How important is packaging to your customer? If you are selling a high-end gift, fashion or luxury item, it may be important to your customer and poor packaging may reflect badly on your offering. For all products, the packaging should be considered part of product design and will affect your cost to manufacturer. Fragile items may need protective packaging. Packaging design can be deferred to later iterations of the design so that you know the size, shape, and weight of the things you are packaging.

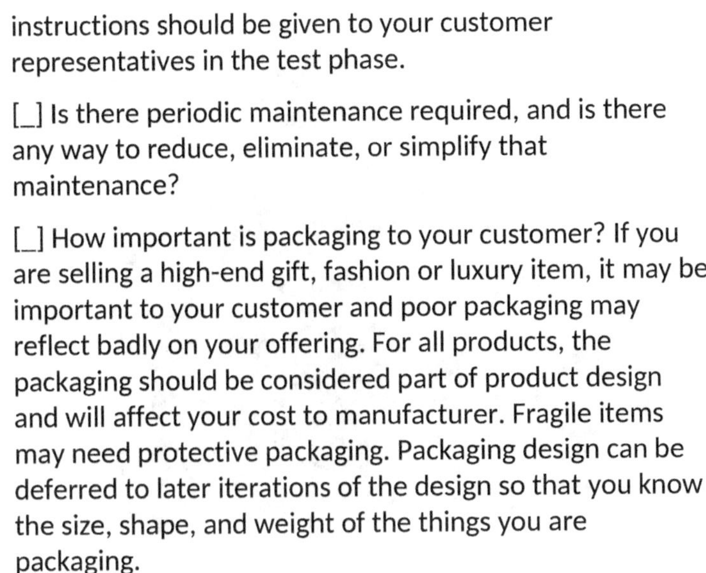

I also want to highlight item 4 in Dieter Ram's list: Make it understandable. Is there any way to have the design inform the user how to operate it? This can be text, graphics, and/or simply the shape of a feature. The more universally you can communicate the better. Not everyone speaks English or will understand graphical symbols. Is there a convention for color coding, or the way something operates that people are used to already? Let their natural assumptions be correct. How can you make it completely intuitive to operate?

Here is an example of poor design in a home yogurt maker that lets you program the time and the temperature. It works quite well, but learning how to program it is confusing because the labels don't match how it actually works. The photo below is of the control panel.

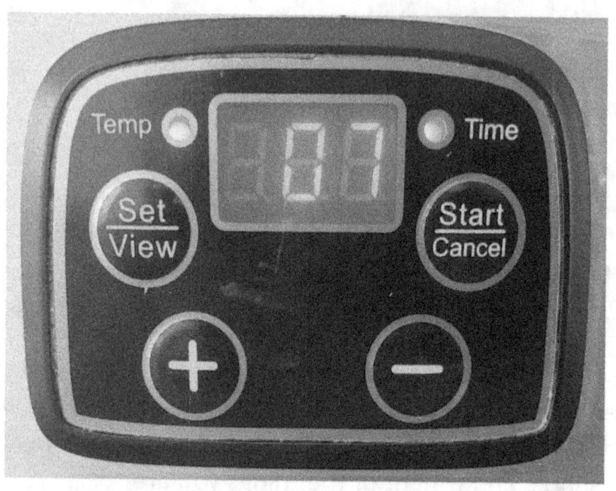

The "Set/View" button allows you to **set** the temperature using the "+" and "-" buttons below it, and this same button will function to **view** the status when the cycle has already started. Similarly, the "Start/Cancel" button will function to **cancel** the cycle if it has already started, but the **start** function is misnamed.

To begin a cycle, you first press the "Set" button and set the desired temperature using the "+" and "-" buttons. Once you set the temperature, it gives you a few seconds to program the time in hours. Can you guess how that's done? You have to push the "Start/Cancel" button, then it will display the number of hours (defaults to 8 hours), and you then press the "+" and/or "-" buttons to change the time. Once the temperature and time are entered, it will start automatically.

The "Start/Cancel" button would be better labeled "Enter/Cancel", since hitting that button during programming will not start the process, but simply **enters** the temperature and then switches to setting the time in hours.

Use international symbols when possible

Consider using international symbols for common functions. IE.:

power functions like on, off & standby

https://en.wikipedia.org/wiki/Power_symbol

media functions like play, pause, fast forward, etc.

https://Wikipedia/wiki/Media_control_symbols.

Making your product easy and intuitive to understand will increase customer satisfaction and may even lower your customer support costs.

Prototypes

In order to get accurate feedback from your customer representatives to refine your design, you will need to create functional prototypes. Even a crude prototype will be better than nothing at all. Being able to touch and feel your creation is not only a thrill, but things also often look different in the flesh than they look rendered on a screen. It's easy for things to seem right on paper, but be wrong in reality, so making it real can help you have confidence to move forward. If you can create functional prototypes that are good enough to try out with customers, you are at a distinct advantage. So how does one go about creating prototypes? Here are some methods for building prototypes for your product.

Your Product Designer may be able to help you create prototypes. As a product designer, I have two 3D printers, and a host of other tools, like clay, paint, etc. At Nala Lab, we are unique in that we are product developers and not just designers. Drop us a line if you think we can help you. Find out more at www.NalaLab.com.

There are 3D printing services that can print in plastic or metal that anyone can access online and have parts delivered by mail. See if you can find a local service that can do any printing you might need. You can also use big service bureaus like Quickparts (https://www.3dsystems.com/quickparts) which can give you an online quote once you upload your 3D part files. I have used them, and they were expensive, but the quality was excellent.

A welder, machinist or woodworker could be employed. It's good to have a network of people and businesses that can help you with this project, and possibly future projects.

In Longmont, Colorado, there is a maker space called "Tinkermill." This maker space is a resource for training and 3D printing, and also has a lot of knowledgeable people. Perhaps you have a makerspace near you.

Modeling clay can be used to facilitate the communication of design ideas with a product designer. Before I learned how to do CAD myself, I would mockup some parts using modeling clay so that the designer could easily get a sense of what I was thinking. A picture is worth a thousand words, and a 3D model in clay is even better. These clay 'prototypes' were of course non-functional, and actually pretty ugly, but proved to be quite useful.

Test and Feedback

This is a critical step, and often skipped by novice entrepreneurs. Getting feedback from others not involved in the development project keeps the project on track and focused on what really matters to your customers. There are always a lot of tradeoffs when developing a product. A cool new feature may add cost (or size, or weight, or complexity, or fragility, etc.) – will customers like the tradeoffs involved in adding this new feature? It's hard to know for sure without testing. But is this important? You already know what the market needs, right? Maybe, but we are prone to cognitive biases that can end up costing us lots of money when we are wrong.

Here are some of the common cognitive traps that can arise in this sub-phase:

Thinking you are smarter than the market: The smarter you are, the more likely you are to fall into this trap. You've probably used your intelligence to take some shortcuts in life, and it usually works out just fine. Your rational capabilities have helped you in many areas of your life. But we need to understand that intelligence is a two-edged sword. That same rational capacity that can help you see things faster and more easily that others can also be used to rationalize to yourself why you are right. Other people would need to do the background homework, but you can just sail through based on your intuition . . . right? I don't think so, at least when it comes to product development.

This is easier to see in other people than it is in ourselves, so consider some of the smart people you know, and ask yourself if they have become convinced of some things that you think are a little off. Do you believe that such illusions only apply to others and not yourself? If true, you are an extremely rare individual. The consequences of being a little wrong (or having some illusions) in most areas of life have an impact, but it is often minor; we can plow along just fine without challenging them. But as an entrepreneur, any illusions you have about your product, or your ability to produce your product, can have long reaching and expensive consequences. The market doesn't care if your product succeeds.

You might be willing to bet a few hundred dollars that you are right, but would you bet $50,000? I ask that, because that's what you're doing if you skip this step. The best way to ensure the success of your product is to verify your assumptions about the market with customers. Trust (your intuition) but verify. When you do, you'll feel much more confident about taking the next steps forward, where there is even more risk.

Falling in love with your idea: It's exciting to create products, and if you didn't love it – at least a little bit – you would probably not be spending all the time, effort and money to bring it into the world. So, falling in love with your idea is probably necessary, but it's not necessary to let this interfere with going through all the prudent steps to verify your design with customers. Instead, fall in love with serving your customers. *The only path to success involves pleasing a lot of customers.*

Stacking the deck: It's tempting to get feedback from one's friends and family, and it's okay to do that, but not as your only source of feedback. Friends and family usually want you to succeed and want to say things they think you want to hear. They are actually not doing you any favors by avoiding critical feedback. You really need that to keep a tight focus on the customer, but friends and family probably don't understand that. Listen to their feedback, and incorporate it as appropriate, but make sure you are also getting feedback from people that don't care if you succeed or fail, and people who are genuinely within your target market. Sure, your grandma loves your new snowboard design, but she's probably not in your market.

Depending on the emotional dynamic you have with friends and family, it's also possible that those close to you will feel jealous or otherwise want to sabotage your efforts. Maybe they feel you should just resign yourself to the kind of life that they have and don't really want you to succeed. If you are not getting critical but constructive feedback, it should not be taken too seriously in most cases.

Letting fear, anxiety, or excitement drive you forward too fast: You're excited by your product, you think it's going to be a big hit, you want and need it to happen quickly because you're running out of money, or time off work, or your partner is losing patience with all the hours you're spending on it, or a thousand other reasons. The temptation is to skip over this rather unimportant step; after all, you're just double checking your assumptions, there is no value added, right? What would a seasoned product developer do?

Customers will perceive your product in ways you have never thought of, both good and bad. Negative feedback can help you improve the product, and positive feedback will help you know how to market your product. Notice the words they use to describe the things they like – these may be the best words to use in your marketing copy. Customer feedback can also help you see applications of your product that you haven't considered. I know someone who sells a fishing vest and that's how it's marketed, but some customers were using it as a photography vest, which expands its market. It's okay to skip over feedback from actual customers for the first few times through the Iterative Development loop, but don't exit from this phase without getting true customer feedback that indicates you have the design nailed.

I hope I've convinced you to do everything you can to get customer feedback before you go into production. It can be challenging to set the testing up and it may stretch you in ways you are not used to. You will need to think a lot about who your customers are and how to find them. Then you will need to reach out in some way to put the product in front of them. In some cases, you can offer functional prototypes for free in exchange for their opinion. If you are concerned about giving away the product but never getting the feedback, you can sell them the product, then give them back the full amount upon receipt of the requested feedback.

How to find customers for testing

You may be able to find customers through user groups, Facebook groups, magazines, blogs, etc. Maybe your friends and family know people who are in the market for your product. When I was younger and was really into cars, I saw some cool-looking wheels advertised on the back pages of a car magazine. I liked them enough to call them up to inquire about getting a set, and they told me they were just gauging customer interest and were not available yet. I'm not sure I'd recommend that approach, but it was a creative way to get customer feedback. Some people have used Google Ads that way as well, testing customer response to certain keywords.

Consider creating a product evaluation questionnaire to give to your reviewers. It's a lot easier to tabulate and summarize the feedback if it is presented in a standardized form. But don't let the questionnaire get in the way of listening to what they have to say – some feedback will not fit easily into a standard format, and often this is the most valuable feedback of all.

My preferred method is to have a survey or questionnaire, but give it to them orally, and listen carefully to their feedback. With their permission, you can record the conversation, so you don't have to rely on memory. The questionnaire helps me to remember to ask all the questions, but also leaves me free to explore anything that comes up that might not be on the questionnaire. Be curious about what drives your customers and be willing to learn from them. A sample questionnaire is included in Appendix B.

Please note that the information you get in this phase not only helps you to refine your product offering, if done well, it can also contain a lot of good information about who your customers are and how your product might fit into their life. Note their choice of words as they describe a problem or their preferred solution. Even if you had a perfect understanding of the market needs, getting this kind of feedback really grounds your marketing materials within your customers' world view.

Revise ideas

The feedback you get may generate ideas about how to improve your design as well as help you better understand your market. Keep asking yourself how you can make the product better and improve the value proposition for your customer. If you have a team, get together to go over all the feedback and brainstorm how your design could be better, as well as sharing any new insights into the market and how to reach your customers. Notice what words your customers use to describe what they want. At a minimum, the Product Champion and Product Designer should review the feedback and determine next steps.

The questions listed previously in the design sub-phase should be asked again here since feedback may shed some light into these areas. Consider adding some related questions to your customer survey to further explore how customers experience your product.

How do we know we are done?

There's an old adage about engineers that if you leave it up to them to decide when to stop improving a product and put it on the market, they never will – they will always have some little improvements or new features to add. This may be true in software development, and perhaps in some company cultures, but for small entrepreneurial business, I think the opposite danger is more prevalent: declaring it done too soon.

The Product Champion (or whoever is paying for the development) must decide when there have been enough design iterations, and that the product is ready to go into production. There is danger in releasing it too soon, and there is danger in holding back for too long. The party or parties taking the financial risk are the only ones whose vote counts as far as I am concerned.

A concept that may help us gauge when the product is ready is known as the Minimum Viable Product (MVP). This is a version of a product that has the minimum feature set required to satisfy the market. If your product at its current iteration has reached the MVP threshold, then it is worth considering whether it is time to go into production. Just because it's in production does not mean that you can't keep learning and preparing for an improved version down the road. The customer feedback we received while in the Iterative Product Development phase is important, but we will learn even more once the product is selling since we will get truly independent customer feedback and a lot more of it than we've had earlier.

The MVP concept grew out of Silicone Valleys' software-centric culture. Keep in mind that the software costs are primarily in the design phase and the cost to "manufacture" software is essentially zero, so software should be biased toward quick release; this may not apply to your physical product. If your tooling costs are low, then it may make sense to release a product with a minimum feature set. If the tooling costs are high, it might make sense to build the full feature set into your product from the beginning. Since I often work with plastic injection molds, I tend to bias toward delaying the decision to begin making the expensive injection mold as long as possible. I like to see a full month or more where I use the prototypes and cannot think of any improvements before pulling the trigger.

It's clearly not time to take the next step if there are quality problems. For example, if the product breaks too easily, or has any safety issues, it's not ready. I've found you can never overestimate how much abuse customers will throw at your product, so make sure it is robust enough to handle rough usage. If the original design intent is not met, then it's not ready. If your target customers don't love it, it's not ready.

We might consider this phase done when we cannot figure out any more improvements. Or it might be when we have a Minimum Viable Product with all the essential features, it's working reliably and safely, and the new ideas for improving it are additional features that may or may not increase the value to the customer. It certainly requires that it is fulfilling the desires of the target customers.

Before we send our design files to suppliers, we may want to consider how to protect our ideas legally.

Intellectual Property

Non-disclosure Agreements: You can create some protection for your product ideas by using non-disclosure agreements with any outside parties that you work with. There are many free versions that can be found on the internet. I will always get a non-disclosure agreement before working with a supplier. They are accustomed to that and it's usually no problem to have one signed. You should use a non-disclosure agreement for anyone you work with, including anyone designing or making prototypes.

You can probably find a free template for a non-disclosure agreement using a simple internet search. I had a lawyer create one for me and it was several pages long and very difficult to understand. Instead of using such a complicated form, I chose instead to write a simple easy to understand non-disclosure agreement. It was all on one page, and I have never had one violated.

Patents: There are three basic types of patents: utility patents, design patents, and plant patents. If you want to patent a plant, you will need to go elsewhere as I am not knowledgeable about them and will not be covering them here.

I'm not a lawyer, and nothing here should be regarded as legal advice. I just want to give you a broad overview of the patent landscape to get you oriented. For legal advice, please consult an attorney.

A utility patent covers the utility of your invention: What it does and how it does it. If you are granted a utility patent, its legal protections last for 20 years.

A design patent is ornamental and covers the look and shape of your product. The classic glass Coke bottle is not something which could be covered under a utility patent, but the shape could be covered under a design patent. If you are granted a design patent, the legal protections last for 14 years.

Once a patent expires, anyone can use that idea without infringing on that patent.

There are 4 criteria that must be met for your invention to be patentable:

> **Subject Matter Eligible:** The invention must be statutory, or subject matter eligible. The patent act states that processes, machines, articles of manufacture, and compositions of matter are patentable. Pretty much any physical device will meet this first criterion. Software generally doesn't, but the way to legally claim a software patent is to claim the algorithms in conjunction with a computing device.

> **Novelty:** An invention must be considered new or novel. One of the first things a patent attorney will do is look for "prior art," or previous patents that cover the same subject matter. But the novelty requirement covers more than just prior patents, it also covers public disclosure. An invention will not normally be patentable if the invention was known to the public or if it was described in a printed

publication before the application was filed. There is an exception to the above two requirements for disclosures made by the inventor (that's you) during the first year before the patent application is filed. This one-year period is non-negotiable, and an inventor has only 365 days after any public disclosure to file an application; failure to do so will make the invention unpatentable. This is the main reason I always get a non-disclosure agreement before telling people about any invention. If your disclosure is covered by a non-disclosure agreement, then that is not considered a public disclosure. So, the non-disclosure agreement doesn't just bind the entity signing it, it also allows disclosure without starting that one-year clock. If there is any chance that you will file a patent application, then using non-disclosure agreements is a must.

Useful: Your invention must be useful or have a purpose. This is generally easily met.

Non-obvious: Your invention must be non-obvious. This is interpreted as not being obvious to a person having ordinary skill in the art to which the claimed invention pertains. For example, changing the size of an invention is an obvious change, so if the only difference with prior art is that it is a different size, then your invention is not patentable.

Utility Patents: You might be wondering if and when you should be applying for a utility patent. I'm not a lawyer, and every situation is different, but for most products, I think the time to file for a patent (either provisional or the full patent) is after we exit the Iterative Product Development phase, maybe after manufacturing, and definitely before the product hits the market. In some cases, it may be better to file earlier in the process, but when you do that, you run the risk of the Iterative Product Development phase giving you design changes that would best be incorporated into the patent application. If you add important features to your product that are not disclosed in the patent filing, then you will either go unprotected for those features or you will have to file additional patents to cover the improvements. Given the high cost of legal help, I find it best to wait as long as possible before filing.

I think most inventors are a little too secretive and paranoid about someone stealing their ideas. One thing I've realized is that the old saw about 1% inspiration and 99% perspiration is way off, it's more like 0.1% inspiration and 99.9% perspiration. This makes me a lot less afraid of sharing my product ideas with people since I know most don't have the skills, money, or time to take my idea and turn it into a product. That said, if I was developing an innovative running shoe, I would not tell any employee of Nike, their families, or friends of their families.

I have seven utility patents. Patents are expensive and time consuming, but they are one of the few defenses a small entity has against the deep pockets and low ethics of many corporations. Many corporations feel no shame in copying your idea if it is not patented but may think twice if it presents legal risk to them.

Patents are also needed if you are planning on licensing your invention. Most companies will not even talk to you if you don't have a patent, and the unscrupulous can more easily steal it when there's no patent. When I was trying to license one of my bicycle products, I sent out a couple dozen letters to companies in the bicycle industry. This was at a time that I had filed a patent, but it had not been issued yet. I did not get much response from my letter campaign, but two giants in the industry did respond: Shimano and Nike. I sent crude prototypes to both, and neither responded with interest. However, Nike came out with a similar product about a year later. By that time, I had gone ahead and manufactured my invention (called Bio-Cleat, a rotating bicycle shoe cleat for clip-less pedal systems). The Nike product was not well designed, and they took it off the market after a year or so; I may have sold more units than they did. I relate this story to let you know that companies will sometimes steal ideas, and that without the passion and insight of the Product Champion, they may not do very well, even if they do have deep pockets.

You should also know that a patent is only indirect protection. It doesn't stop someone from using your idea – in fact the patent document itself tells someone exactly how your product works. What a patent does is give you the right to sue someone for infringement. The possibility of being sued is what provides the deterrent effect, but what are the chances I could have sued Nike and won? With my shallow pockets, almost zero. They could bury me in court costs and hire the best lawyers to argue me out of the game. Even though I found Nike's actions unethical, I never seriously contemplated suing them.

Provisional Patents: If you decide you want to file a patent, you should know about the possibility of filing a provisional patent. A provisional patent is a document that you file with the US Patent and Trademark Office (USPTO) that describes your idea. It is not a formal patent application, in fact, the USPTO does not do anything with your Provisional application other than put it in a file. When you file a provisional, you have one year in which to file a formal utility patent application for that invention. During that year, you have the legal right to claim "Patent Pending" on your product packaging and you will get the benefit of the earlier filing date. It is relatively inexpensive to file a provisional, and in fact I have done it on my own without legal assistance.

I like using provisional patents and I try to file them just before the product hits the market. That gives me a year to assess sales and whether it is worth the time and expense to file a formal patent application and the "patent pending" lets competitors know that they might be sued if they copy my idea. If the sales warrant it, then hopefully I have the cash flow from product sales to support the high cost of a patent. And if sales are not strong, then it may not be worth protecting, and I can save myself the costs.

If you are on a shoestring budget or have some legal skills, you may be able to file a patent yourself. I recommend reading David Pressman's book "Patent It Yourself" to help you get started. It's actually a good resource even if you plan to hire a patent lawyer.

When I write a provisional patent application, I do my best to construct a full patent application with background, drawings, description of my preferred embodiment, etc. It's quite a lot of work, and that work takes time. I used David Pressman's book to educate me on the process. Though I file the application just before sales begin, I have it ready to file way before that time. Otherwise, I may be held up in my product release just to do that legal work.

A few months before the 1-year clock of the provisional runs out,

I give my provisional application to a patent attorney to formalize it and correct any errors that I might have put into it. My work helps them understand the invention with less input from me and may save legal costs. Not all attorneys are willing to work that way.

That's my strategy, and it may or may not be the best strategy for you. I am not your legal counsel, and it may be best to consult yours before deciding on a course of action.

Trademarks: Trademarks are another form of intellectual property, and in many cases, it is easy and inexpensive to file for one. A trademark protects your trade name (brand name or product name), or in some cases, a logo. If the United States Patent and Trademark Office (USPTO) grants your application, your trademark is considered Registered, and that gives you the exclusive right to use your trademark and use the "R" with a circle around it on your packaging and marketing materials representing a registered trademark.

I've received two trademarks from the USPTO by filing them myself for a cost of $225 each. The USPTO will take six months or more to review your application, so plan ahead if you want it ready when your product goes on sale. This is important if you want the registered trademark symbol on your packaging.

Another benefit of having a registered trademark is that when selling on Amazon, owning a trademark allows you into their brand registry program, and that program allows you some important tools like enhanced brand content in your listings, and your own Amazon storefront (currently at no cost).

You can use the USPTO website to search for your intended trademark to see if it is already in use. https://www.uspto.gov/trademarks-application-process/search-trademark-database. If there is no conflict, then the application process is pretty easy. See Appendix A for a step-by-step guide to filing for a simple word trademark with no conflicting use.

Besides filing for a trademark in the US, I have also chosen to file for a trademark in China, even though I don't sell there. The reason I chose to do that is I kept hearing stories about "trademark squatters" – people in China who will take a successful product that's manufactured in China with a US only trademark, and file for that trademark in China. Then they will hold up any exports from China since they are the legal trademark owner and hold that company's shipment for ransom. It's pretty sleazy, but it apparently has happened, and I wanted to make sure it didn't happen to me. It cost me $250 each trademark to file for a Chinese trademark using Legal Hoop.

Domain Names: I consider domain names to be a form of intellectual property. Many of the best names have been taken, and I have often found it challenging to find a trade name I like that is not already taken. Before naming your business or your product, I suggest getting any domain(s) you want, as well as filing for a trademark. If you don't want to get a trademark, I suggest at least ensuring that no one else is using it. I've known some people who set up a business with a name they love, then received a cease-and-desist letter from the trademark owner. Save yourself that pain by checking the trademark database.

Pre-production Planning

When we decide the design is done, we are in the final "Revise Ideas" sub-phase of the (hopefully) final iteration of the product development cycle. Here we do all of our pre-production planning. We need to nail down all of our suppliers, all of the costs, contract terms, and refine our estimates. We should have all the details figured out before committing to production. In some cases, feedback from suppliers will cause another design iteration. The next step involves committing contractually to our suppliers to pay them to produce a product meeting our specifications. It's a big step, and we want to have as few unknowns left as possible before we make that commitment. If your product is something that you can manufacture yourself then have a clear estimate of the time and materials required. In most cases, it's in the production phase that costs and risks will skyrocket, so this is our final chance to be sure we have all the information we need to make the go/no go decision.

Choose your suppliers

So now we are done with the designing and testing and are ready to birth our product into the world, but how do we find the right partner(s)? Maybe you work in the industry and already know some candidates; if so, you are ahead of the game. If you're like most inventors, then you will have to dig in and do some research.

There are many strategies and resources you can use. A critical decision to make is whether to use suppliers in North America or to use overseas suppliers. There are advantages and disadvantages to each:

Language Barrier: It's easier to communicate if you share a language. The manufacturers I have dealt with in China have spoken English, but there was still some lack of clarify and confusion that I don't think would have been there with native English speakers.

Time Zone Difference: Consider how much communication needs to happen in a live conversation vs. via email. If there is a lot of live conversation, time zone differences can be challenging.

Perceived Quality: Many people believe that "Made in the USA" means higher quality. That is probably true in some cases, but there are high- and low-quality companies all over the planet. Still, for some products this should be a consideration.

Shipping Time and Shipping Cost: The farther away your manufacturing is, the more time and money it is likely to cost to get it where you want it.

Respect for Intellectual Property: Some countries may be better than others, but I suspect the biggest differences are from company to company.

Payment Security and Legal Recourse: If there are issues with your manufacturer not fulfilling their legal responsibilities, it will be much easier to deal with if they are located in the same legal jurisdiction as your company.

Made in North America: This might have a marketing advantage for some products.

Lower Production Costs: Production costs are often cheaper overseas.

There may be a larger range of manufacturing capabilities overseas as well. Some things are just easier and more available overseas.

One Stop Shop: One thing I have found is that if you source from suppliers in the US and Canada, you will end up doing the assembly and packaging yourself.

Example: you have your widget manufactured from one supplier, and another supplier makes the paperboard box it is packaged in. A third supplier provides the mounting screws required. Now you have to put together those three items, either yourself or using contract labor. If you need shipping boxes, that is another manufacturer.

When I have done business in China, it's a one stop shop. They provide the manufacturing of all components (or source them for you), packaging, assembly, shipping boxes, etc. It greatly simplifies the process, which is especially important for a small company.

Specifying Critical Dimensions

There may be some critical dimensions in your designs to allow parts to fit together, or for your product to interface with other products. These should be part of the specification that is sent to your suppliers. Your product designer can help you document these.

Specifying Colors

This is normally done using the Pantone system. You can see an online version of the color charts at www.pantone.com, but computer monitors are notorious for having inaccurate color rendering. I recommend getting a Pantone color chart unless you don't care about color and/or are only using black & white. You can order a set online.

Having some color choices can enhance the marketability of your product and allow customers to have choices that suit their style. Color choices also have a downside though, at least for online sales. For some customers, when they are unsure which color they like best, they order two or more colors, keep the one they like and return the others. Getting people to take the risk of online sales generally requires liberal return policies and having multiple colors can work against you in that way.

It can also be challenging for us to know what colors customers will best respond to. I had a friend who has a knack for that, so I paid her to help me with color selection. It can also be good to ask and get feedback from your customers in the test and evaluation phase; showing them the specific colors you are considering is way better than asking them generic questions about what colors they would prefer. If you are on social media, you could also set up a poll to see what colors people prefer.

Another challenge of having multiple color choices is that it complicates inventory management, and generally adds to production costs. Even if you tested color choices in your customer evaluation, once it gets to the market, you may find that the colors that people actually buy don't match up to your initial research. These can also change with the seasons and over the years. You'll run out of some colors before others and have to decide on when to reorder each from your supplier(s).

If you can get away with one color variation at first, I suggest doing that. Once you have a product that is selling well, you can add more colors. This will lower your risk and simplify your life. The same arguments apply to other product variations.

Specifying Surface Finish

Many parts will allow you to select the surface finish. This is a degree of polish or texture on the part. The system that is used is generally SPI. I mostly use SPI B2 for plastic parts but will make my logo/brand name SPI-A2 if it is molded into the part; I think the more polished logo will then stand out better against the background. The surface finishes are listed in the table below. You can ask your supplier to show you examples of the different surface finishes to help your decision. Please refer to the SPI chart below for a general description of each finish level.

Mold Polishing

SPI Finish	Description	Polish Level	um (Min > Max)
A-1	3 micron diamond paste	Optical	0 > 0.025
A-2	6 micron diamond paste	High Polish	0.025 > 0.05
A-3	15 micron diamond paste	High Polish	0.05 > 0.075
B-1	600 Grit sandpaper	Medium Polish	0.05 > 0.075
B-2	400 Grit sandpaper	Medium Polish	0.1 > 0.125
B-3	320 Grit sandpaper	Med - Low Polish	0.225 > 0.25
C-1	600 Stone grinding wheel	Low Polish	0.25 > 0.3
C-2	400 Stone grinding wheel	Low Polish	0.625 > 0.7
C-3	320 Stone grinding wheel	Low Polish	0.95 > 1.05
D-1	#11 Dry Blast Glass Bead	Satin Finish	0.25 > 0.3
D-2	Dry Blast #240 Oxide	Dull Finish	0.65 > 0.8
D-3	Dry Blast #24 Oxide	Dull Finish	4.75 > 5.75

Supplier Directories

Here are some directories that may help you find suppliers:

Thomas Net: North American suppliers from a broad range of industries. Create a free account to be able to search for companies that have the manufacturing capabilities you seek. You can narrow your results by geographic area (i.e., only see suppliers in your home state).

Maker's Row: I've never used them, but you can find American manufacturing capabilities here for apparel and accessory products.

MFG.com: This is one of the easier ways to find a manufacturer, especially if you are looking to manufacture in China and don't have any connections there. They allow you to upload your design files and companies in the appropriate industries can digitally sign a non-disclosure agreement to see those files, then provide a bid. When I did that, I got about eight quotes for an injection molding project I was doing. I followed up with all of them, read the reviews from other customers, and chose one based on both price and quality of communication. I ended up choosing the second lowest quote. It was amazing to me how large of a range the price quotes were - the highest was 5x the lowest! MFG allows the manufacturers to find you, so it is much less work. There is no cost to you – it's paid for with manufacturer fees. Both manufacturers and companies uploading RFQ (Requests For Quote) have a reputation to maintain, which helps insure trust.

Kompass: B2B matchmaking service, matching up suppliers with the businesses they serve.

Haizol: They bill themselves as a global platform connecting international buyers with suppliers in Asia. They are similar to MFG.com in that they allow you to upload a design file and manufacturers will provide bids on your design. I used them successfully for a set of EPDM seals for one of my water bottle lids. Once again, the variation from highest price quote to lowest was quite large.

Alibaba: Alibaba has 58% of the ecommerce market in China. They are also a main resource for Amazon sellers in the US who want to find products from China to sell in the US; it's what I call the "source and sell" model. Many of the manufacturers you find here are also willing to make custom parts, so consider it a starting point. It's a bit of the wild west, but there is a rating system on Alibaba, as well as Alibaba Trade Assurance, so do your due diligence. I have only used Alibaba for small accessory parts, but many Amazon sellers use this as a guide to sourcing complete products.

AliExpress is the consumer retail arm of Alibaba and is geared toward smaller quantity buys.

India Mart: Products from India. Everything from textiles to electronics to industrial machinery.

Internet Search: Using a search engine with keywords tailored to your specific needs can be useful.

Local Library: Libraries are an underappreciated asset. Talk to your librarian to see how they might be able to help you. Some libraries pay monthly subscription fees to have access to online business directories.

Professional Networks: See if you have any connections

on LinkedIn or elsewhere in your professional network related to the industry that represents your product. You can send them a message and ask for referrals.

Automated Injection Mold Quotes from ICO Mold

The company ICO Mold (www.icomold.com) has an online instant quote system. You simply upload your 3D design file(s) and get an instant quote on the tooling cost and product costs based on your design and the materials selected. Their business model is to build molds and do production in Asia but with a strong English language interface to customers. They can also build a mold in Asia and ship it to your local manufacturer.

I have often used their quotes to compare against other quotes I get from potential suppliers. They can also do low volume Urethane casting and CNC machining. So far, I have not used them for a mold, but I have used their instant quoting system many times, especially in the initial High-Level Evaluation phase to get ballpark estimates. By having their online quoting system quote a part or parts with and without a given feature, you can estimate how a feature will affect your startup and production costs.

You can use ICO Mold's quotes for your detailed estimates during the Iterative Product Development phase as well, but the final estimates in that phase should be based on the actual supplier you have chosen.

Contracts and terms

If you are new to entrepreneurship, negotiating contracts can be both intimidating and confusing. I want to give you an orientation to the kinds of terms that you can expect and should be familiar with, as well as some of the strategies used in this phase that can help you control costs.

BATNA: I don't recommend approaching a single supplier. Instead, identify several suppliers and approach all of them at the same time. This allows you to cross-check quotes and leaves you with a Best Alternative to Negotiated Agreement (BATNA). BATNA is a principle from the classic negotiating book "Getting to Yes." Without a BATNA, your negotiating power is diminished. When you approach multiple suppliers at once, each is an alternative to the other – and if one supplier is way off in price or if they are trying to give you difficult terms, you have the others to fall back on. Consider how you would feel in a negotiation with a new supplier if you already had a supplier lined up that could do the job at a competitive price – that is the power of a BATNA.

Approaching Suppliers: I recommend putting together a simple letter introducing yourself and letting them know you are looking for a high-quality supplier within the industry for a new product, and that their company came up in your research. Give them a vague description of your product, but don't disclose any of your intellectual property, whether it is covered by a patent or not. Ask if they are interested in providing a quote, and let them know that, if interested, you will send them a Non-disclosure Agreement (also called a Confidentiality Agreement). The tone should indicate that you are moving forward with the project with or without them, and if they want a chance to play, they will sign the NDA and give you a competitive quote.

The exercise of composing your letter to suppliers gives you an opportunity to think through all the considerations for what that supplier will potentially be providing and puts it all down in one place. It also saves you time so that you don't have to explain the details again to each new supplier you approach. This letter can be updated as you learn more about your specific needs.

RFQ: After a supplier signs your NDA, you must send them your design files as part of your Request for Quote (RFQ). Along with your design files, you should provide drawings with any dimensional tolerances that are required to meet your performance criteria. I like to send a document that describes the project in detail: including materials, surface finishes and color. If there is any assembly being done by the supplier put together a short document specifying the assembly procedure. If a potential supplier thinks your request may be the start of an ongoing business relationship, that will provide incentive for them to send you free samples or otherwise treat you well, so in your Request For Quote, be sure to talk up your company and the potential business you might bring their way.

MOQ: Most suppliers will have a Minimum Order Quantity (MOQ). Even for parts in which no new tooling is required, they may have an MOQ, or they may have quantity discounts that make it cost prohibitive to buy a small amount. You can ask your supplier what their MOQ is, and also ask them for price breaks at various quantities. Obviously, getting your costs down is a good thing and buying more can reduce your costs per unit, but it also increases your initial outlay. If you need a manufactured part in order to complete a prototype, it might be better to buy a small quantity at a higher price to verify that the part works as intended before a major purchase. See also the discussion on samples below.

Samples: If your potential supplier manufactures parts that are similar to what you will be producing, you can ask for samples. Often these will be free of charge, though generally you will have to pay for shipping, which is much more than the product in most cases. Samples can also help you assess the quality of the product your supplier produces.

Tooling and Product Prices: Generally, if you are producing a new product of your own design, there will be some tooling required to get it into mass production. This can include jigs, dies, molds, etc. These costs will affect your project startup costs, and also delay profitability.

These costs are non-recurring costs however, so once they are paid off, they won't affect your production cost. You will need to be aware of your total startup costs as well as your ongoing production costs. These costs are crucial for understanding when and how profitable you can be, and this understanding is crucial before making the decision to go into mass production.

Communication: During all of the back and forth where you are communicating with potential suppliers, make note of how responsive each supplier is. How quickly do they respond to your queries? Did they carefully read your emails, and how clear is their response?

Once you choose a supplier, you may be working with them for years. It is important that there be good communication. This involves good language skills if you are dealing with a non-English speaker, as well as a good attitude and being responsive to your communications. If they are overseas, you won't be able to hop in the car and drive over to pay them a visit.

When dealing with non-English speakers, I suggest communicating with pictures or photos as much as possible, along with arrows and other graphics. Pictures are a universal language and are a good idea for English speakers too.

One Chinese supplier I worked with referred to a plastic injection mold as a "mould" and another spelled it the way I expected: "mold." This was not because the first supplier was spelling it incorrectly; they were using "British English," not "American English." Be aware that international companies are dealing with people from all over the planet and there may be differences like this that arise from time to time.

Lead Times: This refers to how much time it will take your supplier from the time you place an order until it is ready for shipment. I have had four sets of water bottle lid injection molds created in China, plus a mold for a silicone sleeve for a glass water bottle. These were done with two different companies, one of which was found via MFG.com. Except for the silicone sleeve, all of these projects took significantly longer to get done than I was quoted. It is typical for them to quote 45 days for an injection mold, but I have had these projects take close to a year, and in one case, over a year. These schedule delays have major lost opportunity and time to market costs. The price quote, however, has always been honored.

Whatever lead times you are quoted, don't forget to factor in any shipping or assembly required. Also, if you are coordinating parts from multiple suppliers you can try to schedule the deliveries so that the delivery date works out best for your assembly and packaging needs but have some contingency plans in place in case one or more of your suppliers are late. It is particularly hard for suppliers to provide accurate lead times for tooling since so many things can go wrong or cause delays.

Suppliers are generally more accurate when it comes to production times. In the long run, if you plan on placing orders for multiple production runs, you will need to have a good handle on your lead times for each supplier so you can plan to maintain your inventory at adequate levels. I tend to be more comfortable having a buffer to make sure I don't run out, but with a mature product, you can move toward a just-in-time delivery method to keep your warehousing costs down.

Shipping Terms

The shipping terms (Incoterms) define who pays for what parts of the shipment. There is a total of 11 Incoterms (International Commercial Terms – published by the International Chamber of Commerce). Normally shipping quotes are made using FOB (Free On Board) the nearest shipping port. This means they will pack the goods for export, put it on a truck, drive it to the shipping port and have it loaded onto the ship. The buyer pays for the insurance, sea freight, and any charges after it arrives, including customs duties and transportation to the buyer's preferred destination.

I was once quoted with EXW (Ex Works), and I didn't notice because I was expecting FOB. With EXW, the only thing the seller does is package it for shipment. I was responsible for loading it on a truck, driving it to the port, paying the terminal charges and getting it onto the boat. My mistake, and I had to pay through the nose when it came time to get the shipment onto a boat. You are of course free to request your quote in any terms you prefer. Whatever the shipping terms, make sure you know what they are and what your total cost will be. The table below gives a brief summary of the Incoterms and who pays for each shipping charge:

Incoterm:	Export Packing	Inland Loading Charges	Inland Freight	Port Terminal Charges	Insurance	Loading on Vessel	Sea Freight	Port Arrival Charges	Duties & Taxes	Delivered to Destination
EXW - Ex Works										
FCA - Free Carrier										
FAS - Free Along Ship										
FOB - Free on Board of vessel										
CFR - Cost & Freight										
CIF - Cost, Insurance & Freight										
CPT - Carriage Paid To										
CIP - Carriage & Insurance Paid To										
DAT - Delivered At Terminal										
DAP - Delivered At Place										
DDP - Delivered Duty Paid										

Buyer Pays
Seller Pays

Shipping from China generally takes about 5 – 6 weeks to the west coast of the USA. This time can be affected by market conditions; when there is more demand, it can take longer.

Payment Terms

Most of the projects I have done had terms like this:

> 50% of the tooling costs to begin work on the tooling

> 50% of the tooling cost upon approval of the tool's samples

> 50% of the production costs when production begins

> 50% of the production costs when production is complete and before it ships

In practice, the second half of the tooling costs and the first half of the production costs are paid at the same time. The terms of your projects may be different. Just be sure to understand the terms and/or request the kinds of terms that are most suitable for your company. This is another reason I prefer to get quotes from multiple suppliers.

Inspections

You can arrange for inspections of the production run before providing payment. If you are doing production in China, there is a whole industry for inspections, and the prices are reasonable. To find some options, do an internet search for "product inspections (city name)" or "quality control (city name)" where (city name) is the name of the city where your supplier is located. Of course, you are free to travel to your manufacturer's site and do any inspections yourself.

Update your estimates

With the suppliers identified and with firm price quotes in hand, it's time to update the cost estimates from our High-Level Evaluation, as well as to refine the plan for sales and marketing. Be sure to include:

[_] The expected production cost including shipping

[_] All tooling costs and estimates for the development timeline

[_] Production schedule

[_] Warehousing plan and costs for your product

[_] The distribution channels and price structure within each channel: sale price, commissions, shipping charges, fees, etc.

[_] Analysis of your market position and product benefits

[_] Updated market size

[_] How to reach and sell to potential customers and what it will cost

[_] Any other costs associated with getting your product to customers

[_] Overhead costs

Can you start selling without going into mass production?

In some cases, you may be able to do some low volume manufacturing to establish your market. One of my clients used commercial 3D printing services combined with an off the shelf product and his own labor for assembly. This approach allowed him to begin selling his invention and test the market. His unit production costs were higher because 3D printing is generally a lot more expensive than mass production methods, but the financial risk is greatly reduced since there is no or minimal tooling costs and lower volume production runs. It can be a wonderful learning opportunity to have actual market feedback – even better than can be achieved through the Iterative Product Development phase since you will be dealing with actual paying customers.

Hold 'em or Fold 'em Act 2

With all of our updated estimates and exacts costs from all our suppliers, it's time to ask again if the project should move forward. You have a design that's been vetted with customers. You have updated cost estimates with quotes from your suppliers. You've refined your distribution channels, price points, and have a plan for how to reach customers. Your production costs should be well defined, as well as sales prices within your distribution channels. The next step is a big one, and this is your last good opportunity to step away without taking a huge financial risk.

Consider carefully before you move into the next phase of mass production. You might want to take a pause to stand back and reflect before making a decision. You can use the following checklist to help ensure you are ready to move into mass production.

Pre-production Checklist:

[_] Develop a good understanding of how customers will likely perceive your product based on actual customers using your prototypes.

[_] Develop a good understanding of your products market size & how customers are buying similar products.

[_] Ensure the design of the product is complete, having gone through multiple design iterations and feedback from customers indicated that the design is ready for market. The questions starting on page 31 have been addressed if needed. All design parameters are defined (colors, surface finishes, materials), etc.

[_] Establish a detailed estimate of the cost to manufacture, assemble, package and ship your product to your distribution channels. These are based on quotes from the chosen supplier(s) and include all elements of the product including packaging and shipping materials.

[_] Obtain production quotes including an estimate of delivery date for each item. These dates are integrated into a coherent production schedule to support a defined launch date.

[_] Establish design elements such as packaging design artwork and any needed instructions.

[_] Create a plan for warehousing and shipping your product to buyers.

[_] Define initial distribution channels are defined, and any associated costs are identified in detail. If relationships with the initial distribution channels are needed, they are either already in place or there is a plan for establishing those relationships.

[_] Establish a target sales price within each distribution channel based on competitive analysis of other products.

[_] Create sales projections within each distribution channel including costs and estimated profits with at least a pessimistic and an optimistic sales assumption.

[_] Develop a good understanding of your products market size and how customers are buying similar products.

[_] Estimate other costs associated with getting your product out and supported are defined.

[_] Create a marketing plan for reaching potential customers including an advertising budget. If there will be a product launch, this plan is on track to be ready by product launch.

[_] Ensure a strategy for any intellectual property protection is in place (patents, trademarks, etc.).

[_] Obtain all domain names needed for marketing.

[_] Make sure funding mechanisms for getting from production to sales are in place.

Phase 3: Mass Production

Taking the big leap

When you reach this point, it's time to send your final design files and initial financial payments to your suppliers to get things rolling.

Unless you are involved in the actual production, there is little to do for the production but answer your supplier's questions, track their progress, and wait. However, this is an ideal time to prepare for sales. You want to be ready to hit the ground running once the product shows up at your warehouse (or garage!). Much of the following list should already have been thought through in earlier stages and may be partially done. Now is the time to prepare the details.

[_] If you need photos of the production product, take photos of your prototypes as placeholders. Later you can just drop in production photos into the same marketing materials. If you have good production samples these may be suitable for your marketing photos. Even if using older prototypes, there is a lot you can learn about product placement, angles, lighting, etc.

[_] Develop any website(s) that are part of your sales plan.

[_] Write all the copy for your website(s) and sales and marketing materials.

[_] Create any instructions or instructional videos that go with your product. If you can't complete them yet, at least do a storyboard.

[_] If you are going to sell on Amazon, create your listing with the best photos available. Do all of your keyword research, create your title and bullet points. If you have a trademark, you are eligible to be brand registered, so you can create your Enhanced Brand Content.

[_] Create a detailed plan for all of your advertising. If using Google Ad words, do your keyword research and figure out your ads, including any split testing you want to do.

[_] Identify any influencers you want to share your product with and their contact information. Write a custom letter to each and have it ready to go.

[_] Apply for any trademarks if you have not already done so.

[_] If you are planning on filing for a patent, prepare a patent application with the help of a patent attorney if needed. You will want to file an application (provisional or formal) before your product hits the market in order to meet the novelty requirement. If there is a substantial lead time for the tooling and production, consider waiting until your production run has shipped to you before filing. See the previous chapter on intellectual property.

Phase 4: Selling and Support

"Build a better mousetrap, and the world will beat a path to your door."

You've likely heard that one before, but it's only true if people can easily find your door among the millions of other doors, see that your product is better, are convinced your product offers something that they genuinely want or need, it's offered it at a price they find compelling, and obtaining it is not too much work. Many great products have failed because of marketing failures. Connecting buyers and sellers is not automatic or trivial and can be a major challenge to do well.

When I was doing engineering for a large company, I didn't really appreciate the role of people in sales and marketing. I thought engineers were somehow superior since they understand how things actually work – or at least how the products actually work. Now having had many marketing failures due to not understanding the bigger picture of how the product fits into the marketplace and what it takes to have it gain traction in that marketplace, I have a lot more respect for those professions. Understanding how you will reach your market once your product is available should be part of your product development plan.

All of the previous phases led to us having a product to offer which we think nails the market need. We live in a world where there is intense competition for buyers' attention, and if you want to sell lots of product, you'll need to:

Get people's attention

Convince them of your product's value

Give them an easy path to obtain it

That last point is the easiest to understand. You have to somehow get your product into the customers' hands, and that means you need at least one distribution channel.

Product Distribution Channels

There are various distribution channels available for your product and you should understand how they work, your costs, price, and the obligations for each. This should have been defined as far back as the high-level evaluation, but by the time you get to this phase, you should have refined and/or validated your ideas about product distribution. Many if not most products use multiple distribution channels. Let's take a look at the main channels.

Online: For online sales; you can sell on your own website, Amazon, Ebay, Etsy, Walmart and/or others. Be sure to understand your shipping options and costs and include packaging and shipping costs in your profitability calculations. International shipping of course costs more.

The online option has the lowest barriers to entry, and

generally low overhead. But this does not make it easy to sell there. I have sold on Amazon for over a decade, and it has become harder and harder to make any money there. Amazon's fees have increased, and competition has continued to grow, stressing profit margins.

Amazon policies are great for the customers, but not so great for the 3rd party sellers. For example, if I sell using the Fulfilled By Amazon (FBA) program that means Amazon has my product in their warehouse and is responsible for sending out customer orders. If a customer complains that an FBA shipment was late, Amazon will often refund the customer's money from my account. They get points with the customer, and I pay the price, even though they were the ones shipping late. Amazon also fosters price competition (good for the customer, bad for the sellers), and has a program to recruit Chinese sellers, most of which are selling rebranded versions of the same copycat product, and their main selling point is low price, which creates a race to the bottom on prices. This happens because Amazon's main criteria for displaying a product to a customer's search is its sales velocity. The more you sell the more you are seen by customers and low prices attract more sales. It's a difficult environment to compete in.

Wholesale & **Distributors:** Getting into retail stores may be more difficult but is at least simple in concept. Many retailers will not want to buy direct from you since they probably sell many different products, and managing too many suppliers increases their business costs. In such cases they would likely prefer to purchase your product from a distributor. Distributors warehouse many products from many manufacturers so they can consolidate product shipments to their retailers.

You can generally charge for shipping cases of product to a retailer or distributor but be aware of the markups to your product price that happen along the way. For a low-cost product, say $10 or less, a retailer will generally mark it up by 100% (double the price), and a distributor will generally add 30%. Exact markups will depend on the price and industry, so please investigate the mark-ups in your product segment.

Let's take an example product that you sell to a distributor for $4. If the distributor marks it up 30%, they will sell it to the retailer for $5.20, and the retailer will turn around and sell it for $10.40. In order for you to make any money, you will need to be able to produce the product for an all -n cost of around $2 or less. For a quick estimate when evaluating the viability of a potential product, I want to see it retail for 5x or more of its manufacturing cost.

Distributors and retailers are bombarded with requests from manufacturers to carry their products. You will have to approach them and convince them that they can make money by selling your product.

Direct Retail Sales: For some products, it might make sense to sell directly to customers yourself by setting up a retail storefront. At a low level this might be setting up a booth at a farmers' market or flea market. This kind of boutique selling has the advantage of being low cost and low risk to set up, and the further advantage of giving you direct interactions with your customers, from which you might learn a lot. Even if you plan on nationwide or worldwide distribution, starting super small like this might help you refine the product, instructions, or packaging in order to make your wider distribution run smoother. Remember, fixing mistakes early in the process will save a lot of time and money in the long run.

Of course, a permanent retail location might also involve a brick-and-mortar retail location, or even a chain of them. This obviously has high overhead costs for a building and staff, so it would only be appropriate for certain projects, and would have a high startup cost.

Marketing

One of the most challenging parts of product development for me is getting people's attention and convincing them my product is something they want and need. The great temptation here is to shout out how great my company's products are. If you take that approach, you will be in competition with millions of other voices, each vying for the customer's attention. People learn to tune it out.

To get your customer's attention, you need to understand who your customers are and what they want. One effective approach I have come across is Donald Miller's Story Brand framework as outlined in his book "Building a Story Brand – Clarify Your Message So Customers Will Listen." The brand story framework takes a different approach. Instead of making your products the hero of the customer's journey, the brand story framework recognizes that the customer is the one on a hero's journey.

Your company is the trusted guide that gives them a plan for solving the hero's (customer's) problems so that they can fulfill their destiny and avoid disaster. Your customer is Luke Skywalker, and your company is Obi Wan Kano Be who gives Luke a plan to save mankind from the Evil Empire. As you might guess, this framework is based on the classical movie storylines, which are all about capturing our attention and holding it. Consider what problem(s) your product solves, and how having your product could impact their lives.

What are your customer's challenges?

What does your customer want?

What kind of benefits will overcoming these challenges provide to their life?

How will success feel to them?

The main tasks for the marketing phase are to define and implement:

[_] Your sales channels, pricing, and support for each channel.

[_] Your value proposition: what are you offering the customer.

[_] Your customer relationships: how will you connect with your customers? Social media, website, customer support, user groups, etc.

[_] How will you reach new customers? Advertising, marketing, referrals, etc.

I'm not a marketing expert, and that's also not what this book is about, but marketing is an important part of any successful product and unless your business is really more of a hobby, my advice is to take it seriously and plan it in from the start.

As you put together copy for your website, ads, and other marketing materials, try to use the kind of language that your customers use. If you've honed your product offering by getting feedback from customers, you already have some of the language. Make note of what words they use to refer to a feature, as well as how they describe the benefits of your product. Notice what they say about how it has improved their lives. Customer interviews and surveys are very helpful here (see Appendix B).

Another approach I appreciate is that given by Seth Godin in his book "This is Marketing." He says that the days of marketing to the masses is over and that the successful company today is focused on an individual market segment; it creates products that serve that smaller market segment perfectly, rather than trying to create a product that doesn't offend anyone but doesn't excite anyone either. He looks for the smallest viable market and serves that market well. He suggests focusing on permission-based marketing where your customers want to interact with you, and even be involved in the development process. It involves building relationships of trust, and those are the kind of relationships that can last for decades.

Crossing the Chasm

Many products have a period where the innovation is not immediately accepted by the market. This is covered in the classical book "Crossing the Chasm" by Geoffrey A. Moore. The market segments are shown below:

Product Adoption Curve

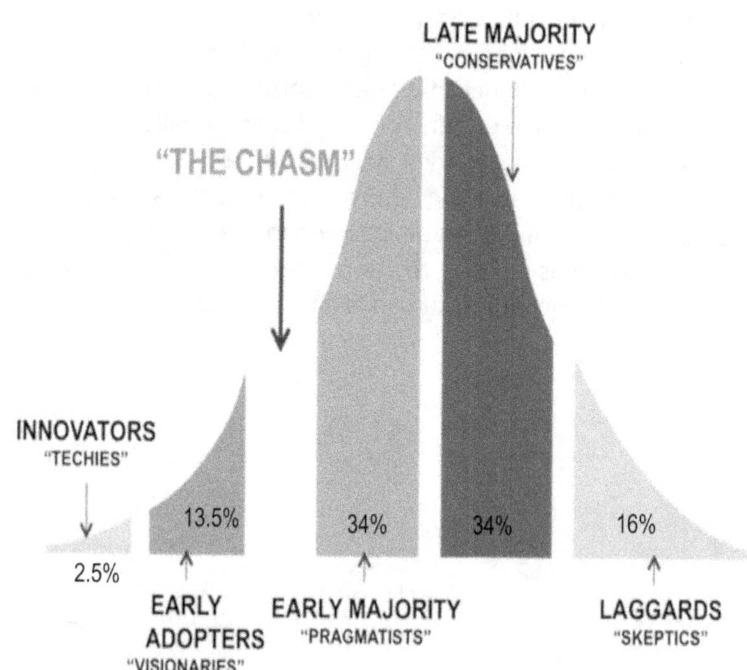

The Chasm is the gap between the early adopters and the early majority. Many products don't make it across that gap to the early majority. This gap is present for products that take a new and different approach to solving a problem, especially one that requires adopting new technology. It could be a business-like Uber or Air BnB but might also simply be a new way of doing things. I helped a client create a nail clipper for babies that had an adjustable aperture to prevent injuring an infant's skin – it looked different and works differently than what people are used to. If your product utilizes a new technology or a new approach to solving a problem, I highly recommend reading the book.

The chasm is the gap between the early adopters and the early majority. The early adopters are enthusiastic about the type of innovation you are providing and are willing to take the risk. They'll forgive you if your product is not perfect, especially if you show willingness to improve it over time.

The early majority does not want to take any risk. They expect the product to have all of the bugs worked out. If they have any problems, you can expect scathing criticism. They don't care how hard you worked or how innovative it is, they just want it to work the way they expect, be reasonably priced, and not have any quality problems. It's just a widget to them that solves a need, and if it creates any hassles in their life, that shows you where the product is not quite ready for prime time (sales to the majority of the market). Many products have trouble crossing the chasm into mainstream acceptance.

The late majority is even more risk averse. They want to see the product proven and accepted by the early majority before they are willing to risk a purchase.

If you want to make money from your product, you'll probably need to find acceptance by the majority – there are just not enough people among the innovators and early adopters to support a thriving business for most products.

Customer Service and Support

Have you ever bought a product, then had a problem or question about it? Maybe there was a missing part, or confusing instructions? If so, you may have reached out for customer support. It's an important part of being in business and you'll need to figure out what kind of support your product needs, as well as how you will provide that support. Having good, easy to understand instructions may help, as may having a good website with answers to commonly asked questions. I send out a user's guide for a simple water bottle straw lid and have that information available on my website as well. I have generally only offered email support for my products, and as long as you are responsive, this may suffice. If you have developed your product with abundant feedback from customers, you probably know how much support your product will need.

Whatever you decide to do, there are costs involved in setting up and supporting your customer service operations. These should be built into your business plan from the start.

Summary

If you have read this far, thank you for investing in your own success. Creating a successful product requires you to have everything lined up and dialed in. I often picture business success being like shooting an arrow through a series of tires hanging from a rope. If one of the tires is too low, too high, or off to one side, it will be difficult or impossible to shoot an arrow through the center of all of them to reach your target customers.

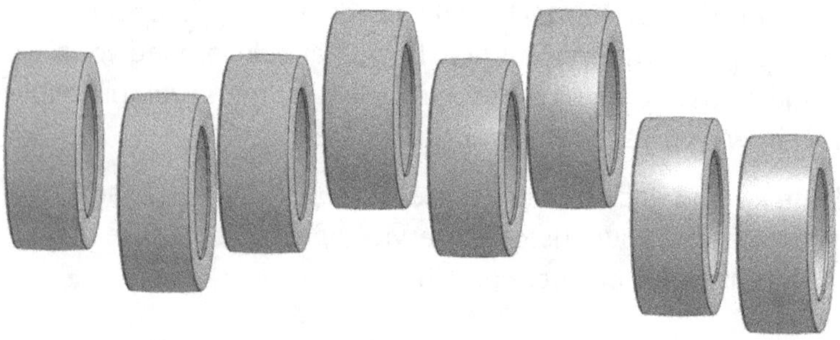

Here are a few of the "tires" that need to be lined up:

[_] You need to understand your customer's true needs and desires and your product needs to satisfy them.

[_] You need to have a clear message to your potential customers that lets them know how your product can assist them.

[_] You need to find a way to get that message to your customers in a way that interests them.

[_] Your product must offer a compelling value to your customers so that they are willing to pay you for it.

[_] You need to have a way for your customers to purchase your product that is easy and convenient for them.

[_] You need to get your product to your customer in a timely way.

[_] Your product needs to be free from unwanted hassles or problems.

[_] You need to offer the kind of information and/or support they need in order to have a positive experience with your product.

If you get all of the above right, you just might find yourself with a successful product. May it be so!

Should you take the risk of bringing your product idea to market?

When considering whether or not to pursue a particular product, I will often use my imagination to walk down two different paths: one where I develop the product and one where I choose not to. When choosing to move forward, usually I can imagine feeling content that I have tried, even if I fail.

When I choose not to, I often feel some disappointment or

regret, like I missed a golden opportunity out of fear. But other times it feels like it's of no consequence, and that gives me important information too. I have had more product ideas than I can count and being able to take only the best of them is part of the art of product development.

Practice, Practice, Practice

I probably would not pay to see someone play the piano if I knew that person had just taken up the piano and never practiced. It's practice that makes us good at something, and product development is no different. If you want to get good at product development, it helps to practice. Each iteration of the Iterative Product Development cycle is like a practice session for a given product, and each new product developed will widen and expand your knowledge and skills. I often practice product development by 3D printing small items around my home. It doesn't matter that much if it could become a viable business, just that it gives you practice in refining your craft. Unless you have money to burn, it's helpful to practice over in the inexpensive end of the pool. Taking a product idea through the high-level idea phase or even through the iterative design phase is a better education than you can get in any school. Because it's so educational, it is not a waste of time or money even if you don't move forward into production and then sales.

Further Reading

I highly recommend the book "Nail It, then Scale It" by Nathan Furr and Paul Ahlstrom. Their approach mirrors the Iterative Product Development method. The iterative design process is where they "Nail it", creating designs that satisfy the customer based on validated feedback. After the product design is nailed, then it gets scaled, which in this book refers to mass production.

Final Words

Hopefully you have read to the end before spending too much money or time investing in your product ideas. I hope this short guide can serve to make the process easier and help you avoid the mistakes that I have made. If you would like any help with design or product evaluation, feel free to reach out to me.

alan@eafproducts.com

www.nalalab.com

Epilogue

There are some additional things I'd like to share about product development that don't fit neatly into the previous chapters. Most of these are philosophical or spiritual in nature, and often apply to the process overall.

One of the most viewed TED talks is one by Simon Sinek called "How great leaders inspire action." In this inspiring video, he explains that "people don't buy what you do, they buy why you do it." You can watch it here:
https://www.ted.com/talks/simon_sinek_how_great_leaders_in spire_action/c

For me, bringing a product to market is both an exciting opportunity for me and is an act of service to my fellow humans. I'm not interested in bringing a product to market unless that product will make my customers lives better in some way. It doesn't have to be a big way; it just has to make some small improvement in their lives. So, my "why" is that I am contributing in some small way to the betterment of the human condition. The specifics of that improvement may form the nucleus of a marketing approach. In any case, it is fulfilling to me to think I made some small difference in people's lives.

Whether it is used in marketing or not, having a clear purpose can help guide the formation of your enterprise, and help align your actions and especially your hiring decisions.

Where do product ideas come from?

The ideas I've had for products seem to come unbidden, as if they are downloaded through some type of human group consciousness. That may sound a little wacky, but I have had the experience during trade shows where I was introducing my new product and have had several people come up to me excitedly and tell me how they had that idea too, but never worked on it. It really seems like an idea comes through the ether into the minds of many people, but only a few will take up the call to bring it to market. Whether we are all connected in that way, or whether there is some other explanation, often products seem to be based on an idea whose time has come.

This probably explains some cases of multiple people developing a similar product and one thinking that the other must of spied on them and stole their idea. It might just be that they both were receptive to the same improvement idea. I have experienced being on the other side of this too where I get to see a product I'd envisioned being introduced by someone else: as long as I have not pursued it, it is invigorating to me.

Mistakes are inevitable

Of course, we want to minimize our mistakes, but they are impossible to avoid completely. What I tried to do in this guide is shift the mistakes to be as early in the development as possible, where errors are much less costly. But don't think they can be avoided completely, and don't beat yourself up if and when it happens. Being an entrepreneur requires being able to pivot when things don't work out as planned.

Appendix A: Applying for a Trademark

Step by step guide for applying for a Trademark from the US Patent and Trademark office (USPTO). This covers word marks only.

Before you begin

[_] Find a trademark that you want to use.

[_] Check to see if the domain name is available (if applicable). You will want to purchase the domain as soon as you file your trademark – don't delay or someone else may take it! There is a rather sick business model that involves others applying for domain names based on trademark and business name filings so that they can sell the domain name to you at an inflated price.

[_] Find or take a photograph of your product that clearly shows the trademark on the item or packaging. You will be uploading this photo during the application process.

[_] Decide what legal entity will own your trademark. For example, this could be owned by you, or it could be owned by your LLC.

[_] Search the USPTO database to see if anyone is using your trademark by first going to https://www.uspto.gov/trademarks-application-process/search-trademark-database

() Select *Trademark Electronic Search System (TESS)*

() Select **Basic Word Mark Search (New User)**
Enter your trademark name. Be sure you are spelling it EXACTLY the way you want to use it, including any spaces, etc. Select **Submit Querry** and note the results. If your mark has never been used, you will get a screen like this:

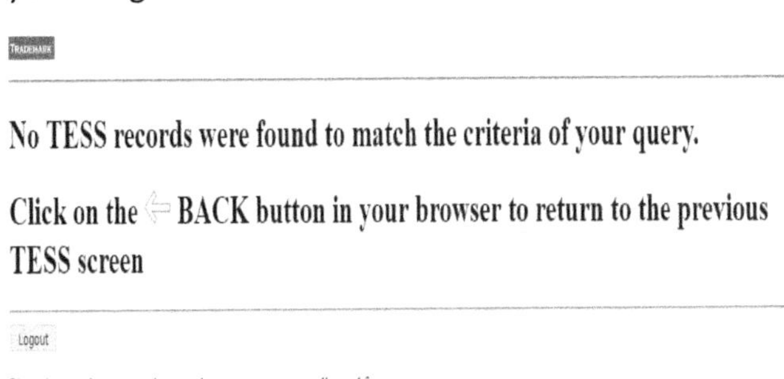

() Find the classification of your trademark using the master list: You should discover ALL uses of your trademark (all possible product categories that the mark will cover). Note the search phrases you used to find the categories – you will use these in the application process.

https://tmidm.uspto.gov/id-master-list-public.html

If you have no conflicts with your trademark, you can proceed to file the application. I'm not sure how the application process

works if your trademark is already in use; it is said to be a bit
more complicated, so you may want to consult with an attorney
for that.

You may want to review the required information so that you are sure you have everything on your computer before starting. Please note the items marked in **bold** in the list below.

FILE USING THE TEAS ONLINE SYSTEM

Go to https://www.uspto.gov/trademarks-application-process/filing-online/initial-application-forms and choose TEAS Plus ($225) or TEAS RF ($275). I recommend TEAS Plus.

Fill out the form. This is the information you will be providing:

[_] Is an attorney filling out this application? **No**

[_] Owner of Mark: **Your name or company name, etc.**

[_] **Entity Type** (Individual, LLC, etc.)

[_] Address, phone number, email, web site

[_] Trademark Name: **YOUR TRADE NAME**

[_] Click "Add Goods/Services", then enter **your search phrase(s)** you found in the master list of categories (Last step in before you begin). Repeat "Add Goods/Services" until complete.

[_] You will use section 1(a) or section 1(b), depending on whether your mark is in use. I suggest waiting until it is in use and then using section 1(a).

[_] Attach a photograph with your product showing your trademark on the product or packaging.

[_] Description of Specimen (the photo you attached).

[_] Date of first use of mark anywhere

[_] Date of first use of mark in commerce

Click on: Assign Filing basis

Check the 4 boxes under declaration.

Fill out the bottom section with your signature, name, position, phone number and date. Your signature is in the format:

"/FirstName LastName/" or

"/FirstName MiddleName LastName/";
in other words, put a slash before and after your name to make it "an electronic signature."

Appendix B: Sample Interview Questionnaire

The following questionnaire has been adapted from the Marketing Map developed by the Paradigm Shift Project. The marketing map provides an easy-to-follow, empathy-based method based for developing your marketing copy in a way that will appeal to your customers. The approach is highly effective for everyone from a marketing novice to those more experienced. The genius behind this is Kai Madrone, so if you reach out, I recommend you arrange to work with her if possible. To find out more about the Marketing Map, visit https://www.theparadigmshiftproject.com/strategic-storytelling.

Please use this as a guide and tailor it to fit your individual product(s). Feel free to add detailed questions that relate to your product. For example, if you come up with a new golf club design, you can ask what golf clubs they are currently using, and what they like and don't like about them.

In regard to XX (the subject area for your product):

A. For getting at the **external problem:**

>What's your situation now regarding XX? What's not going right?

>How is that a problem?

>What's the number one thing you struggle with the most?

B. For getting at **desire/what they want:**

If you could have anything in regard to XX, what is it? What do you most want?

If you had a magic wand and you could wave it and have anything in regard to XX, what do you most want?"

C. For getting at the **internal problem:**

What's the biggest hurt or pain in regard to XX?

What's the worst part of the current situation? How is the lack of a workable solution affecting your life?

D. For getting at what **failure** is for them:

What would be the impact on your life if these problems remained unsolved?

If you don't turn this around now, what would you lose out on?

E. For getting at their **obstacles and objections:**

What is stopping you from getting where you want to go?

What is slowing you down, or standing in your way?

What other obstacles do you experience?

F. For getting at what **success** is for them:

If you could overcome these challenges and flow freely toward your goals, what would that do for you?

What would your life be like?

How might it change you as a person?

How would your relationships change?

How would your life change?

How does success in this area feel in your body? *(this last one gives you their language for how success feels, which can come in super handy for both tone of your copy and images you include in marketing materials)*